Elements of
LAND/SOIL POLLUTION

Elements of
LAND/SOIL POLLUTION

Om Prakash Gupta
B. Tech. (Chemical Engg.)
Ex DGM, HRD
Bokaro Steel Plant
Steel Authority of India Limited (SAIL)

KHANNA BOOK PUBLISHING CO. (P) LTD.
Publisher of Engineering and Computer Books
4C/4344, Ansari Road, Darya Ganj, New Delhi-110002
Phone : 011-23244447-48 **Mobile:** +91-9910909320
E-mail : contact@khannabooks.com
Website : www.khannabooks.com

Price: ₹ 299.00

Elements of Land/Soil Pollution

Om Prakash Gupta

ISBN: 978-93-82609-73-5

Edition: 2019

Published by:

Khanna Book Publishing Co. (P) Ltd.
4C/4344, Ansari Road, Darya Ganj, New Delhi-110 002
Phone: 011-23244447-48 **Mobile:** +91-9910909320
E-mail: contact@khannabooks.com

Printed in India by:
vaibhav Products

PREFACE

It gives me immense pleasure to present this book titled 'Elements of land/soil pollution'. The book contains five chapters followed by appendices pertaining to complete domain of land/ soil pollution. Latest and classified information are given in the appendices. Besides, it also contains two chapters devoted to short questions & answers and multiple choice questions & answers drawn from the examination papers of various engineering colleges for the benefits of the students. Last chapter contains glossary of terms related to soil pollution.

This book is targeted to benefit the following:

- Diploma in engineering students
- Degree in engineering students
- AMIE, AMIIM, AMIICHE Students.
- Faculty members and teaching staff
- Practicing engineers / professionals / Consultants

The scope and coverage of the subject matter presented can be gauged from glancing the contents portion of the book to judge its suitability and relevance for a particular target group. The subject matter has been presented in a simple, lucid and easy to understand manner for easy grasping of the subject.

Latest and updated information/data/statistics pertaining to the subject matter has been included in this edition for the benefit of the readers.

Constructive criticisms and suggestion for further improvement of the book are solicited.

Ranchi, Jharkhand
26th January , 2018

Om Prakash Gupta

Contents

1
CHAPTER
All about Soil— Its Formation, Types, Components & Properties

EARTH AND ITS STRUCTURE

Earth is a large mass detached from sun, when a powerful object collided with sun.

The earth is of size 12757 km in diameter.

The earth is divided into 3 main layers as shown in **figure** namely:

- Core- inner & outer
- Mantle
- Crust

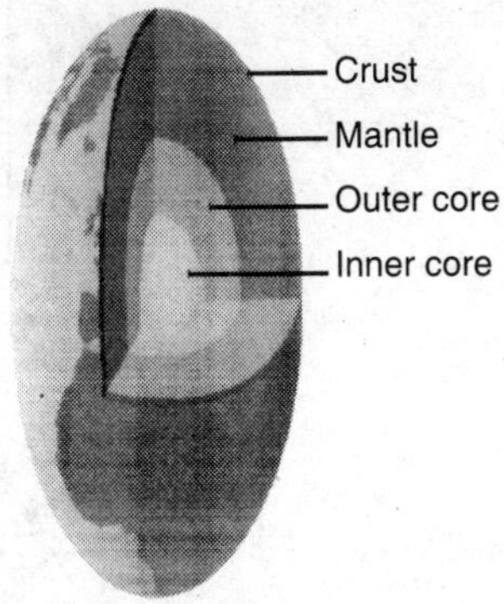

Fig.1.1: Main Layers of the earth

Crust: The outermost layer of the earth is the crust which comprises of varying rocks; the thickness of crust varies from 64 to 96 km and about 70% of the crust is immersed in Oceans. Crust comprises of minerals like silica, aluminium, magnesium.

Mantle: The next layer is the mantle which forms the most of the earth's mass. It is about 2880 km thick. Most of the internal heat of the earth is located in this region. Mantle comprises of minerals like chromium, silicon, magnesium, nickel, iron.

Core: The inner most part of the lithosphere is the core; this layer surrounds the center of the earth and its thickness is about 3500 km. This is hottest, heaviest and densest layer of the earth. Core Comprises of iron & nickel and other components like, sulphur, silica & cobalt.

Lithosphere

- The topmost part of the earth surface (crust)which cover most of the land containing hills, grasslands, deserts is called lithosphere.
- After the formation of earth, the rocks were formed first by cooling of the earth mass and successively the soil was formed due to degradation & fragmentation of the rock due to physical, chemicals & weathering effects.
- The lithosphere includes all the metals, minerals and inorganic & organic matter present in the soil. The minerals are essential for the growth of vegetation and ultimately to the entire living organisms as they are the basic elements in food cycle.

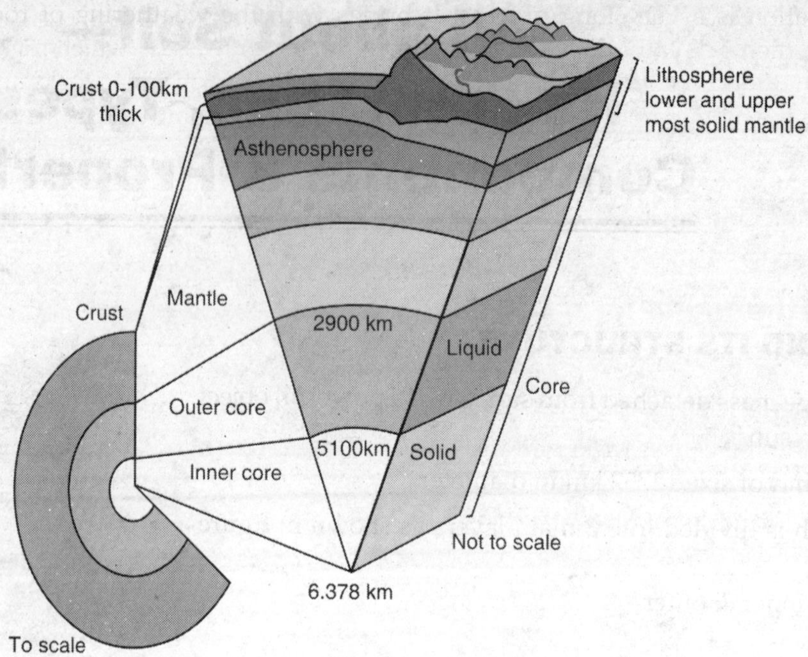

Fig.1.2: Structure of earth

The top layer of earth is called soil. More than 80% of the land is covered by soil and it is 3 to 4 meter in depth on the earth surface. It is one of the most important natural resources. It consist of a mixture of minerals, organic material, living organism, air and water that together support the growth of plant life. The traditional meaning of soil is the natural medium for the growth of land plants. Many terms are used to refer to the non-water surface of the earth. The soil scientist calls it soil, a geologist may call fragmental rock or overburden, an engineer may call land. Land is a general term to describe the non-water portion of the earth, usually in reference to a large surface area.

Mechanism of the formation of soil

Rocks of crust are continuously subjected to mechanical, chemical and biological action. As a result, they are broken down into finer particles during long period of time. The process of breaking up of rocks into finer particles is called as weathering.

Changes in temperature, beating effects of hails, rain water, pressure of winds and steams causes mechanical weathering. Addition or removal of oxygen within rock materials and action of carbonic acids cause chemical weathering. Sometimes, plants like lichens growing in rock crevices produce some chemicals. Those chemicals help in the breaking up of rocks into finer particles. Such type of weathering is called as biological weathering. The process of soil formation is called as Pedogenesis and study of the soil is called Pedology.

The weathered out fine particles get mixed with dead remains of plant & animals and with their excretory materials to from soil. Formation of soil from the parent material (bed rock) due to mechanical weathering of rocks by temperature changes, abrasion, wind, moving water & glaciers and chemicals weathering activities & lichens under ideal climatic conditions may be about 1cm of soil within 15 years from soft parent material.

Soil formation is a very long process. It brings with the weathering of rocks minerals into small fragments. Weathering is the name given to the process by which rocks are broken down through direct contact with the planet's atmosphere to form soils. Rocks and geological sediments are the main parent materials of soil (the materials from which soils have formed). There is a very wide variety of rocks in the world, some acidic, some alkaline, some coarse-textured like sands, and some fine-textured & clayey. Weathering occurs with no movement, and thus should not be confused with erosion, which involves the movement of rocks & minerals by agents such as water, ice, wind, and gravity.

Rocks Weathering to form Soil

There are three main types of weathering; physical, chemical and biological.

Physical Weathering

Physical weathering causes rocks and minerals to disintegrate into places without altering its chemicals composition and it is considered as an isochemical process. Physical weathering is the influence of processes such as freezing & thawing, wetting & drying, and shrinking & swelling on rocks and other sediments, leading to their breakdown into finer particles.

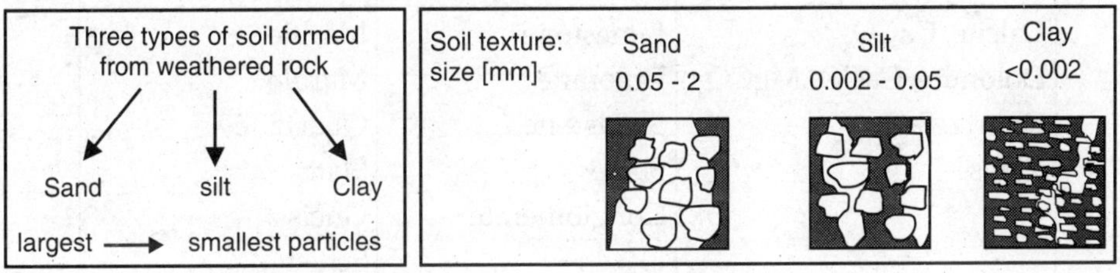

Fig.1.3: Process of Physical Weathering to form soil

Chemical Weathering

Chemical weathering is the decomposition of rocks through a series of chemical processes such as hydration, hydrolysis, carbonation, dissolution and oxidation. Rainwater is slightly acidic due to dissolved carbon dioxide and reacts with calcium carbonates in limestone rocks. This reaction could dissolve limestone slowly.

Biological Weathering

Biological weathering is the effect of living organisms on the breakdown of rock. This involves, for example, the effects of plant roots and soil organisms. Respiration of carbon dioxide by plant roots can lead to the formation of carbonic acid which can chemically attack rocks & sediments and help to turn them into soils.

There are a whole range weathering processes at work near the surface of the soil, acting together to break down rocks and minerals to form soil. These weathering are constant & gradual processes and have given rise to most of the world's soils.

Table 1.1: Primary and Secondary Minerals in soil

Primary Minerals	Secondary Minerals
Quartz, SiO_2	Geothite, FeOOH
Muscovite, $KAl_3Si_3O_{10}(OH)_2$	Hematite, Fe_2O_3
Microcline, $KAISI_3O_8$	Gibbsite, $Al_2O_3.3H_2O$
Orthoclase, $KAISi_3O_8$	Clay minerals (Al silicates)
Biotite, KAI (Mg, Fe)$_3$(AISiO$_3$O$_{10}$) (OH)$_2$	Dolomite ($CaCO_3.MgCO_3$)
Albite, $NaAISi_3O_8$	Calcite ($CaCO_3$)
Hornblende, $Ca_2Al_2Mg_2Fe_3Si_6O_{22}(OH)_2$	Gypsum ($CaSO_4.2H_2O$)
Augite, Ca_2(Al, Fe)$_4$(MgFe)$_4$Si$_6$O$_{24}$	
Anorthite , $CaAl_2Si_2O_8$	
Olivine , (Mg,Fe)$_2$SiO$_4$	

Table 1.2: Dominant Mineral and Rocks in soil

Dominants minerals	Sedimentary rocks	Metamorphic rocks
Calcite, $CaCO_3$	Limestone	Marble
Dolomite, $CaCO_3.MgCO_3$	Dolomite	Marble
Quartz, SiO_2	Sandstone	Quartzite
Clays	Shale	Slate
	Conglomerate	Gneiss
		Schist

Factors Affecting the Soil Formation

Climate; Living organisms; Relief; Parent material; and Time.

Classification of Various Types of Soil

(A) Types of Soil on the Basis of Particle Size

On this basis of size of particles or according to their texture, soil is into three main types:

(a) sandy soil, (b) clayey soil, and (c) loamy soil.

Sandy Soils: A soil having loosely packed coarse or large sized particles is called as sand. Sandy soils are well aerated, warm and are easy to work or dig. This type of soil remains dry most of the time and their nutrients are quickly washed out. This is because, it cannot hold water.

Clayey Soil: It is made up particles of very small sizes. Those particles remain compactly arranged. It is sticky and cohesive or waterlogged when wet. It can hold greater quantity of water.

Loamy Soil: It is made due to mixing of clay (20-30%), silt (30-45%) and sand, which make them fertile & easy to work, is called as loamy soil. A compact soil made up of rather small sized particles as compared to sand is called as silt. Loams are the best soils for plant growth

Sometimes, soils contain dead and decomposed parts of plants & animals. Such a decomposed organic matter found in the soil is called as HUMUS. It is crucial to maintaining the balance of minerals and nutrients needed for plant growth. Presence of humus in the soil is essential for high production of crops. Farmers add humus in the soil by mixing manure (compost, wormy compost or natural manure) to the soil.

(B) Types of Soil on the Basis of Origin

On the basis of origin, soils are classified into following major groups:

- Residual Soil
- Transported Soil
- Movntainous Soil

Residual Soil

The soil which is formed due to weathering of rocks and is found adjacent to those rock is called as residual soil. This soil is basically composed of mineral particles of different sizes.

On the basic of presence of mineral components, the residual soil is further divided into: red soil, Black soil and laterite soil.

Red soil is red in colour due to the presence of oxides of iron in it. Besides iron, it contains potassium in good quantities. The **black** soil is made up of ash and lava that come out during volcanic eruptions. This soil contains aluminium, iron, calcium and magnesium in good amounts but it lacks phosphorus & nitrogen. This soil is suitable for the cultivation of cotton crops. The **laterite** soil is formed due to weathering of open parts of rocks. It is red-yellow in color. It contains hydrogenous oxides of iron and aluminium. This is usually acidic hence unfit for the growth of plants.

Transported Soil

The soil which is transported through gravity, flowing water, wind or glacier from one place and deposited to some other place is called as transported soil. This soil has further been put into four different categories on the basis of mode of its transportation. Those four different categories are colluvial, Alluvial, Sandy and Glacial or drift soils.

Colluvial soil is formed due to deposit of soil in low areas due to land-slides.

Alluvial soil is formed due to deposits of particles contained in the water of current. This soil often contains different layer and this is principally composed of fine-rounded particles.

The soil which is made of sand particles deposited by wind at a certain place is called as **sandy or desert soil**. Only desert plants(xerophytes) grow in this type of soil.

Some soils are formed due to deposits of particles through glaciers. Those soils are called as **glacial soils**.

Mountain Soil

The soil which is found on mountains, hill slopes, cliffs and valleys, is called as mountain soil. This type of soil is basically composed of pebbles, clay shell and limestone. The lower part of this soil is usually calcareous in nature. This soil is found in north-east Himalayan regions of India.

Soil and its Components

Soil is, naturally occurring, unconsolidated or loose covering of broken rock particles and decaying organic matter on the surface of earth. It is capable of supporting life. Where there is no soil, there is no plant life and we have barren rock and/or sand. Soil is therefore of critical importance to life. Simply put: no soil, no life. Luckily for us, soil covers most of elements required for nutrition of higher plants.

The Soil Composition

Soils are made up mainly of four components.

(i) Minerals (ii) Organic matter (iii) Water (iv) Air

In most soils, minerals represent around 45% of the total volume, water & air about 25% each and organic matter from 2% to 5%. The composition is graphically represented in figure below.

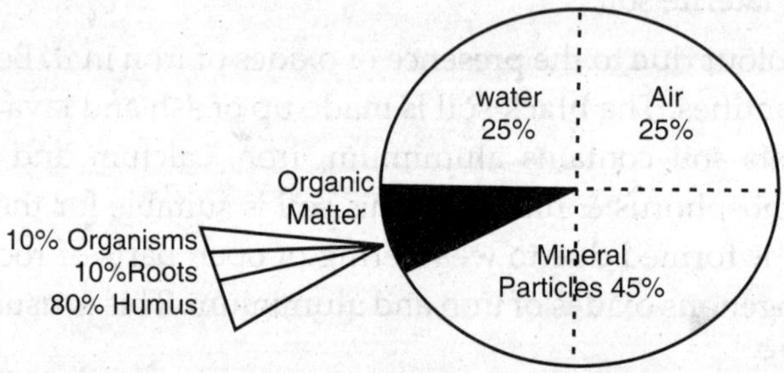

Fig.1.4: Composition of an average soil

Average Composition of Soil by Volume

Inorganic Mineral- 45%; Organic matter-5% Pore space-50% (25% Air & 25% Water).

There are five components of soil: Rock, sand, silt, clay and humus.

- **Rock:** A rock is a solid and indefinite mixture of naturally occurring substance, mainly minerals. It is inorganic and naturally formed substance without a particular atomic structure of chemical composition. Rocks are composed of one or more minerals or one or more elements. Rocks may range in size from tiny pebbles to huge mountains. Minerals are naturally occurring solids with a definite chemical composition and crystal structure. They are simply a collection of one or more elements that are stacked neatly together in a form called a crystal structure. Minerals are also solid and formed naturally in the earth.

  Soil Size: Soil size is an important physical property and classed into three size range: Sand, Silt and clay. Other particles larger than 2 mm are not classified as soil particles, and are categorized as gravels, stones, cobbles or boulders. Large size of particles such as stones and gravels have no contribution to soil properties. Gravels: 2-4mm; Pebbles: 4-64 mm; Cobbles:64-256 mm and Boulders:> 256 mm in size.

- **Sand** is the largest soil particle, ranging in size from 0.05 mm to 2mm, and is sub categorized within that range to include coarse, medium & fine sand. It does not hold water or have many nutrients. Sand is a naturally occurring granular material composed of finely divided rock and mineral particles. It is actually a mixture of; silica (SiO_2), Calcium Silicate ($CaSiO_4$), Calcium Nitride (Ca_3N_2), Silicon Nitride (Si_3N_4), Aluminum Nitride (AIN_3), Alumina (Al_2O_3), Borazone "Boron Nitride" (BN), Magnesium Oxide (MgO), Silicon Oxysulfide (SiOS), Lithium Silicate (Li_2SiO_4), and other oxides/nitrides of a multitude of metals. Sand is not a single source compound but a mixture of a lot of organic and inorganic minerals, naturally occurring in nature. But it's mostly composed of silica (quartz-SiO_2). It is relatively inert; results in little effect on soil chemistry & pH, poor nutrient holding capacity (i.e. CEC), less total pore space, larger (macro) pores, fewer small (capillary) pores; thus sand causes: increased aeration, increased drainage and decreased water holding capacity.

The term "sand" is applied to loose granular materials falling within a specified particle size range. The widely used terms are silica sand, quartz sand & industrial sand, but in fact they are quartz rich sand and used for a variety of industrial material or process as well as for construction purpose, since they are with a very low or negligible silt, clay and organic matter content.

- **Silt** is the next smallest particle size, which ranges in size from 0.05 mm down to 0.002 mm. It is made up of rock and mineral particles that are large than clay but smaller than sand. Individual silt particles are so small that they are difficult to see. Silt is found in soil, along with other types of sediment such as clay, sand, and gravel. Silty soil is slippery when wet, not grainy or rocky. Silt can change landscapes. For example, silt settles in still water. So deposits of silt slowly fill in places like wetlands and lakes. Floods deposits silt and settles in still along river banks and on flood plains. Deltas develop where rivers deposit silt as they empty into.
- **Clay** Particles are the smallest soil fraction, less than 0.002 mm in size. Clay is a soft, loose and earthly substance containing mineral group Feldspar, known as the mother of clay. It has sheet like structures and are composed of mainly silicate and aluminate groups. Chemical formula of feldspar is Al_2O_3, $2K_2O$, $2SiO_2$. Pure Clay is known as kaolin and has the formula $2SiO_2$, Al_2O_3, $2H_2O$. Clays are divided into three classes: **porcelain Clay**, which is approximately pure Kaolin. This burns to white or light-cream color; **Plastic clay**-which contains more impurities than the porcelain clay. It burns to a yellow-red color and is used for ordinary earth ware; **Fire clay**-these clays approach very close to the porcelain clays in composition, however they contain a large quantity influence on soil-water relations, determining water holding capacity, air & water movement, and also plant nutrient retention & exchange. It holds water well. It is sticky and can be shaped when it is wet. But, it is very hard when dry. Clay has many nutrients.
- **Humus** is made of leaves, twigs, small animals, or other decayed substances. Humus adds many nutrients to the soil. Humus is present in the top soil.

Soil Horizon or Soil Profile

Since soil is a natural & three dimensional body, the surface view of soil reveals very little about the depth or its subsoil characteristics. Therefore, to get view of the soil, we must expose a vertical section. A soil profile is a view of a vertical cross section of soil from the ground surface downwards to where soil meets the underlying rock. Soil is made up of distinct horizontal layers; these layers are called horizons. As we move deeper the rocky component becomes predominant, and at a certain point the rock becomes homogeneous & compact; this is the mother rock.

Soil horizon thickness is an essential parameter for plant growth and environmental issues (soil erosion etc.). Geomorphologic position and topography (such as elevation, slope and hydrological & erosion processes) influence the depth of horizons & soil properties. Therefore, soil horizon depth varies across a region. The farm management, land use, or risk management need these information for decision support.

Soil horizontal layers are called horizons. They range from rich organic upper layers (humus and top soil) to underlying rocky layers (sub soil, regolith and bed rock). A vertical slice through the different horizons is called soil profile as shown in figure below.

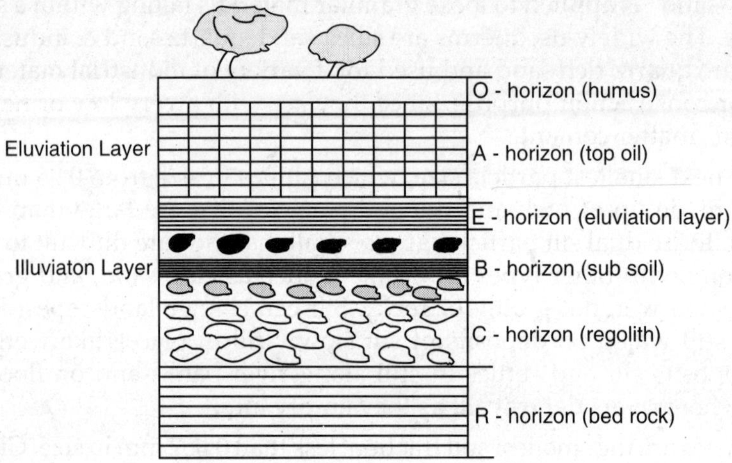

Fig.1.5: Soil horizon or soil profile

The profile tells us a great deal about the factors that interact in the formation of a soil. Feature of various soil layers/ horizons are described below.

O-Horizon: The top organic layer of soil, made up mostly of dead organic matter (detritus) deposited by plants-leaves, stems, fruit seeds etc., and humus (decomposed organic matter). Its color is brown or black. O-Horizon is high in organic content and is the primary energy source for the soil ecosystem. Towards the bottom of the horizon, the process of decomposition is well advanced and the original materials may be unrecognizable. At this point the material is dark and is called humus.

A-Horizon: This layer called top soil. It is a mixture of mineral soil from below and humus from above. Seeds germinate and plant roots grow in this dark colored layer.

***O&A-* Horizon:** It contains large amount of bacteria, fungi, earthworms, small insects; forms complex food web in soil, recycle soil nutrients & contributes. to soil fertility.

***E-* Horizon:** E-stand for eluviation, which refers to the process of leaching of many minerals because of the downward movement of water It is made up mostly of sand and silt. Having lost most of its minerals and clay as water drips thought the soil (in the process of eluviation). This layer is more pale in color than the two layers above.

B-Horizon: Also called the sub soil. It contains clay and minerals deposits like iron, aluminum oxides and calcium carbonate. It is yellow or reddish in color. It is strongly influenced by illuviation. It contains less organic material & fewer organisms than A- Horizon.

C-Horizon: Also called regolith. It consists of slightly broken-up bed rock (parent material). Plant roots do not penetrate into this layer; very little organic material is found in this layer. This layer is affected very little by the biological and chemicals processes that go on in the upper layers. It is not significantly influenced by the pedogenic processes & translocation. Chemical composition helps to determine pH of soil & also influences soil's rate of water absorption & retention.

***R*-Horizon:** The unweathered rock layer (bed rock) that is beneath all the other layers. It is the final layer in a typical soil profile.

Significance of soil

- The soil is the natural medium in which plants grow. It provides oxygen, water and minerals to plants-roots.
- It supports life and forms the basis of plant's & animal's life.
- We get food, clothing and building material from the plants & crops that grow in soil.
- Soil is a habitat of numerous types of organisms — plants, animals and home for a number of friendly micro-organisms that maintain & enhance the fertility of the soil.

Soil sustains all life on earth. Roots of plants grow in the soil and absorb nutrients & water which is used during the photosynthesis process, and also for the growth & development of the plants. Consumers, like herbivores and omnivores, eat plants.

Herbivores are eaten by carnivores and omnivores in the food web, so nutrients that are absorbed from the soil will benefit all organisms. Soil provides a habitat for many organisms. When people abuse the land through careless acts, irresponsible behavior, greed and pollution, the soil is harmed and so too all the organisms that directly & indirectly require fertile soil to survive. Forms of soil pollution are litter, dumping of toxic waste, deforestation etc.

Litter is a major problem of modern society because man-made materials, like plastic, glass and tin, are non-biodegradable. Litter results when wastes are not disposed in a regulated and proper manner. Rubbish should be placed in disposal containers for collection and dumped in landfills that are covered with soil. Exposed litter is the breeding ground for rats, mosquitoes & flies, apart from being unsightly, unhygienic and a danger to animals & plants. Toxic chemicals must be disposed off in sealed containers that are buried in special cemented landfills. If these landfills are not managed properly, containers may leak. This will result in streams & rivers becoming contaminated, causing people and animals that drink the water, to be poisoned. Crops may absorb the toxins and when eaten by people, birds & animals, cause diseases and toxic poisoning. Unscrupulous people illegally dump containers of toxic chemicals into the sea where the containers eventually corrode and leak, releasing the toxic waste. This will directly impact on the marine plant and animal life in the area.

Following Corrective action can be put in place to prevent soil pollution

- Educate people to reduce, re-use and recycle wastes by providing containers to recycle paper, glass, tin & plastic.

- Enforce legislation to prevent dumping in unauthorized areas with strict fines.
- Large fines should be levied when people litter.
- Use only indigenous trees for afforestation.
- Educate farmers to ensure crop rotation.
- Provide informal settlements with proper waste removal.
- Ensure local government provides many dustbins in towns and cities.
- Form action groups in communities to clean up polluted areas.

Soil plays significant roles in the cycling of three major greenhouse gases: carbon dioxide, methane, and nitrous oxide. Land management practices that increase the amount of carbon (organic matter) stored in soil are reducing the amount of CO_2 released to the atmosphere. Microorganisms in poorly aerated soils, such as in rice paddies or wetlands, produce methane as they decompose organic matter. Nitrous oxide is a product of denitrification in soil. As well as being a greenhouse gas, nitrous oxide (N_2O) depletes the stratospheric ozone.

Soils have suffered from several decades of damage due to various reason on a scale that we must not allow to continue. There must be stronger awareness among the public, policy makers and users to preserve soils for current and future generations.

Effects of soil Pollution

- Pollution of surface soil, underground soil and disturbances in soil profile;
- Reduced soil fertility;
- Reduced nitrogen fixation;
- Larger loss of soil nutrients;
- Reduced crop yield;
- Imbalance in soil fauna and flora
- Pollution runs off into rivers and kills the fish, plant & other aquatic life;
- Corrosion of foundations of any structure and pipelines;
- It is difficult to remediate contaminated soil sites; and
- Adverse effect on human and animal health through consumption of crops growing on such soils.

Soil health

Farmers often use the term 'soil health', which is similar to the term 'soil quality' used by soil scientist: it needs to incorporate adequate chemical and biological properties: it needs to incorporate adequate organic matter, have a good structure, and be home to a diverse mix of organisms. These properties allow the soil to carry out important functions, and may be achieved in a natural setting by a soil reaching equilibrium with its surroundings, or in managed setting by human intervention to improve the soil's health. Agriculture soil health is linked to human health, as poor soils yield fewer crops with decreased nutritional value. Healthy soils also limit erosion, and help improve air & water quality.

Contamination can seriously affect soil's ability to perform some of its key functions in the ecosystem. Soil is a living resource, but once contamination exceeds a certain threshold, the soil may be considered 'functionally dead' . Pollution by heavy metals and many organic contaminants is practically irreversible.

Soil Quality

Soil quality is the capacity of soils within landscape to sustain biological productivity, maintain environmental quality and promote plant & animal health.

Causes of Soil Quality Degradation

- Overgrazing;
- Deforestation and Soil erosion;
- Agriculture Activities;
- Population and Over exploitation of soil; and
- Urbanization & Industrialization.

Mechanism of soil quality degradation

- Water erosion
- Wind erosion
- Chemical degradation
- Physical degradation

Soil Quality and Functions of Soil

As compared to the environmental component of air & water, soil is an extremely complex system. This is due to a huge variety of soil properties and chemical, physical & biological system condition. The combination of these makes the development of general rules for quantitative evaluation of soil quality impossible. This is because soils have a much higher buffering capacity than air and water.

Soil Quality in Relation to Soil Function

The quality of soil is adversely influenced by contamination (pollution) of the system. Contaminants mean the compounds that may exert adverse effect on soil functioning. It can be defined as soil contamination. Because most soils do have a certain buffering capacity, it usually takes some time before the negative effects become apparent. Once situation occur, the soil can be considered as polluted, which for all practical purpose means that malfunctioning or dysfunctioning of the soil is apparent due to abundant presence or availability of the compounds. Such malfunctioning may refer to one specific function in particular but sometimes also in combination of different function are:

- **The bearing function**: For instance, as playground for children and for building of houses/structures.
- **The Plant growth function:** This may refer to natural vegetation and the production of crops for animal & human consumption.
- **The filtering function of water:** Ground water as well as surface water.
- **The ecological function of soil**: With its contribution to element cycling as an important aspect.

Proper functioning of soil as stated above impose a wide variety of quality criterion. This strongly hampers the introduction of a quality assessment methodology with general applicability and validity. At first sight, the bearing function would probably be the least demanding, at least once certain requirements are met. However, experience has shown that for building of houses, some minimal requirement of chemicals conditions in the soil are needed: and a combination of use the same site for waste disposal and latter for urban expansion is not always without problems.

Method to assess health risks for people living in area where is an unconscious combination of soil usages or in polluted areas, requires insight in the quantitative relationship between pollutant exposure of the organism under consideration on one side and the resulting effects to be expected from the exposure on the other side.

In the case of human, the exposure pathways of soil contaminants are:

- Soil ingestion, especially by children.
- Inhalation of air, containing volatile polluting compounds.
- Drinking water.
- Food in the form of plant products and animal products.
- Radiation exposure

As a result of the adverse effects of heavy metals and other contaminants, environmental agencies proposed critical levels in soil above which toxicity is considered to be possible. The table below shows the normal range and critical total concentration of heavy metals for soil.

Table 1.3: Normal range & critical total concentration of heavy metals in soil

S. No.	Heavy metals	Normal range in soil (mg/kg)	Critical soil total concentration (mg/kg)
1	Cd	0.01-2.0	3-8
2	Co	0.5-6.5	25-50
3	Cr	5-1500	75-100
4	Hg	0.01-0.5	60-125
5	Pb	2-300	100-400
6	Ni	2-750	.100

APPENDIX 1.1

SOIL ACIDITY DESIGNATIONS BY PH VALUE

S. No.	Type of soil	pH value
1	Extremely acidic soil	<4.5
2	Very strongly acidic soil	4.5-5.0
3	Strongly acidic soil	5.1-5.5
4	Medium acidic soil	5.6-6.0
5	Slightly acidic soil	6.1-6.5
6	Neutral soil	6.6-7.3
7	Mildly alkaline soil	7.4-7.8
8	Moderately alkaline soil	7.9-8.4
9	Strongly alkaline soil	8.5-9.0
10	Very strongly alkaline soil	>9.1

Note: Early soil classification were based on individual's characteristics such as the texture of soil (e.g. loam, clay or sand) or the parent material (e.g. alluvial soil, gravelly soil etc.) but a more scientific classification is based on the combinations of soil characteristics in relation to their formation. There are eight major of types of soil according to Indian Council of agriculture Research (ICAR).

1. **Alluvial Soils:** Alluvial soil is formed when a soil-carrying water stream gradually loses its carrying capacity with decreasing velocity. In slowing down, a river does not have sufficient power to keep the large particles of soil suspended; these particles settle to the riverbed. Further decreasing velocity causes smaller particles to settle. Alluvium are material, such as sand, silt or clay, deposited on land by streams. Alluvial soil is transported soil.

   Thus, it is formed by deposition of alluvium by rivers and contributes greatly in agriculture development. It is further classified in two types- 'Khadar' and 'Bhangar'.

   KHADAR- It is old alluvium represents riverine alluvium of Himalayan rivers. It is clayed & dark in colour, coarse in nature, and contains kankar (lime nodules), pebbles & gravels.

   Characteristics - Coarser in the upper section and finest in the lower section; Light to dark in colour; Rich in Potash & Humus; poor in Phosphorous & Nitrogen; Highly fertile and good for all crops (kharif and rabi).

   Suitable for Crops- Rice, Wheat, Sugarcane, Cotton, Jute etc.

2. **Black Soil:** It is created from igneous rocks and also known as Regular or Black Cotton Soil. It is dark grey to black in colour and high in clay content. Highly suitable for cotton crop and rich in iron, lime, calcium, magnesium, carbonates & alumina. It is poor in phosphorous, nitrogen and organic matter. Black soils develop cracks in summer. The soil in black is colour because it is volcanic in origin.

   Suitable for crops- Cotton, Sugarcane, Wheat and Oilseed etc.

3. **Red Soils:** Red soil is formed due to weathering of old crystalline rocks in the areas of low rainfall. The red colour is because of the diffusion of iron in the form of iron oxide. It looks yellow when it is the hydrate form. It is sandier and less clayey. Rich in iron, small amount of humus; poor in phosphorous, nitrogen & lime; slightly acidic and do not retain moisture or water. It is porous and friable.

   Suitable for crops- Tobacco, Potato, Sugarcane, Rice, Wheat.
4. **Laterite Soils:** It is formed under high temperature and rainfall with wet & dry spell. It is brown to yellowish in colour; Silica is leached due to high rainfall; Remnants of iron and aluminum oxides left behind is known as Laterite. It becomes hard when exposed to atmosphere and used as building materials. Rich in iron; poor in lime, potash & magnesium. The humus content in the laterite soil is less because the micro-organisms and decomposers get destroyed in the high temperature.

   Suitable for Crops- Tea, Coffee, Cashew, Rubber and Coconut.
5. **Desert Soils:** Originated by mechanical disintegration and wind deposits. It contains soluble salts, red to brown in color, porous & coarse, 90% sand & 5% clay, rich in nitrate & phosphates, poor in evaporation is faster and soils lacks humus & moisture.

   Suitable for Crops- Drought resistant crop, like millets and barley.
6. **Mountain Soils:** It is found in hills slopes. Formed by deposition of organic matter from forests. Rich in humus; poor in potash & lime.
7. **Saline and Alkaline Soils:** It contains a larger proportion of salts like sodium, magnesium & calcium; infertile and unfit for cultivation. It is sandy to loamy in texture. They have more salt, largely because of dry climate and poor drainage. They occur in arid & semi-arid region and in water logged & swampy areas. Their structure ranges from sandy to loamy. They are also known as 'Usara soils'.
8. **Peaty and Marshy Soils:** Occur in humid region; formed by accumulation of poorly decomposed organic matter; black in colour; highly acidic and heavy. They are rich in free alumina-iron compounds of blue color and poor in lime. The soil become waterlogged during monsoon and are generally very poor for agriculture.

   Soil types are also classified according to many more factors such as colour, depth, productivity, etc.

APPENDIX 1.2

FUNCTIONS OF VARIOUS MACRONUTRIENTS & MICRONUTRIENTS IN SOIL

S. No.	Macronutrients	Function
1	Nitrogen(N)	Nitrogen is a part of all living cells and is a necessary part of all proteins enzymes and metabolic process involved in the photosynthesis & transfer of energy. Nitrogen is a part of chlorophyll, the green pigment of the plant that is responsible for photosynthesis. Helps plants with rapid growth, increasing seed & fruit production and improving the quality of leaf & forage crops. Nitrogen often comes from fertilizer application and from the air (legumes get their N from the atmosphere, water of rainfall contributes very little nitrogen). Plants absorbs nitrogen by absorbing either ammonium or nitrate through its root system. The plant will then utilize nitrogen as building block to produce protein in the form of enzymes.
2	Phosphorus (P)	Like nitrogen, Phosphorus (P) is an essential part of the process of photosynthesis. Involved in the formation of all oils, sugars, starches, etc. Helps with the transformation of solar energy into chemical energy; proper plant maturation; withstanding stress. Effects rapid growth. Encourages blooming and root growth. Phosphorous comes from fertilizer, bone meal, & superphosphate.
3	Potassium (K)	Potassium is absorbed by plant in larger amounts than any other mineral element except nitrogen and, in some cases, calcium. Helps in the building of protein, photosynthesis, fruit quality, reduction of diseases, strong root formation & sturdy steams, full grown crop & having longer shelf life and increases crop yield. Potassium is supplied to plants by soil minerals, organic materials, and fertilizer. It is in the soil naturally in two forms, one form is able to be absorbed into the plant, while the other is unavailable to the plant.
4	Calcium (Ca)	Calcium, an essential part of plant cell wall structure, provides for normal transport and retention of other elements as well as strength in the plant. It is also through to counteract the effect of alkali salts and organic acids within a plant. Sources of calcium are dolomitic lime, gypsum, and superphosphate.

5	Magnesium (Mg)	Magnesium is part of the chlorophyll in all green plants and essential for photosynthesis. It also helps activate many plant enzymes needed for growth. Soil minerals, organic material, fertilizers, and dolomitic limestone are sources of magnesium for plant.
6	Sulphur (S)	Essential plant food for production of protein. Promotes activity and development of enzymes and vitamins. Helps in chlorophyll formation. Improves root growth and seed production. Helps with vigorous plant growth and resistance to cold. Sulphur may be supplied to the soil from rainwater. It is also added in some fertilizers as an impurity, especially the lower grade fertilizers. The use of gypsum also increases soil sulphur levels.
7	**Micronutrients:** Boron (B)	Helps in the use of nutrients and regulates other nutrients. Aids production of sugar and carbohydrates. Essential for seed and fruit development. Sources of boron are organic matter and borax
8	Copper (Cu)	Important for reproductive growth. Aids in root metabolism and helps in the utilization of proteins.
9	Chloride (CI)	Aids plant metabolism. Chloride is found in the soil.
10	Iron (Fe)	Essential for formation of chlorophyll. Sources of iron are the soil, iron sulphate, iron chelate.
11	Manganese (Mn)	Function with enzymes system involved in breakdown of carbohydrates, and nitrogen metabolism. Soil is a source of manganese.
12	Molybdenum (Mo)	Helps in the use of nitrogen. Soil is a source of molybdenum.
13	Zinc (Zn)	Essential for the transformation of carbohydrates. Regulates consumption of sugars. Part of the enzymes systems which regulate plant growth. Sources of zinc are soil, zinc oxide, zinc sulphate, zinc chelate.

Note:

- Essential elements are required for the normal growth of plants;
- Macronutrients are necessary in large amount (usually 50 mg/kg in the plant) for the growth of plant; and
- Micronutrients are necessary in only extremely small amounts (<50 mg/kg in the plant) for the growth of plant.

Sixteen chemical elements are important (and present in soil) to a plant's growth & survival. They are known as plant nutrients. These are divided into two main groups namely non-mineral and mineral nutrients.

The non-mineral nutrients are hydrogen (H), oxygen (O), & carbon (C) and are found in water. They are consumed by plants by photosynthesis process. There are 13 mineral nutrients, which come from the soil, are dissolved in water and absorbed through a plant's roots. The mineral nutrients are divided into two groups: **Macronutrients and Micronutrients.** Macronutrients are divided into two more groups: primary and secondary nutrients. The primary nutrients are nitrogen (N), Phosphorus (P), and potassium (K) . These major nutrients usually are lacking in the soil because plant use them in large amount for their growth and survival. The secondary nutrients are Calcium (Ca), Magnesium (Mg), and sulphur (S). There are usually enough of these nutrients in the soil, so fertilization is not always needed. Also, large amounts of Calcium and Magnesium are added when lime is applied to acidic soils. Sulphur is usually found in sufficient amounts from the slow decomposition of soil organic matter, an important reason for not throwing out grass clipping and leaves.

Micronutrients are those elements essential for plant growth which are needed in only very small (micro) quantities. These elements are sometimes called minor elements or trace elements. The micronutrients are Boron (B), Copper (C), Iron (Fe), Chloride (CI), Manganese (Mn), Molybdenum (Mo), and Zinc (Zn). Recycling organic matter such as grass clipping and tree leaves is an excellent way of providing micronutrients (as well as macronutrients) to growing plants.

Table 1.2T$_1$: Macronutrients and Micronutrients in soil

Macronutrients	Micronutrients from soil solids
Mostly from air and water: • Carbon • Hydrogen • Oxygen **Macronutrients from soil solids:** • Nitrogen • Phosphorus • Potassium • Calcium • Magnesium • Sulphur	Iron Manganese Boron Zinc Copper Chlorine Molybdenum

Table 1.2T$_2$: Reference of soil Macronutrients for agriculture purpose:

Nutrients	Soil Reaction (pII)	Electrical Conductivity (in mmhos/cm)	Organic carbon (%)	Nitrogen (in ppm)	Phosphate (P_2O_5) (in ppm)	Potash (K_2O) (in ppm)
References	6.5-7.5	1.0-2.0	0.5-0.75	280-560	23-56	130-335

Table 1.2T$_3$: Reference of soil micronutrients (in ppm) for agriculture purpose:

Nutrients	B	S	Zn	Cu	Fe	Mn
Reference	0.5-1.00	13-20	0.78-1.20	0.60-1.20	7.00-12.0	3.00-5.00

Typcial Agricultural Soil

An agriculturally productive soil is a balanced mixture of sand, slit and clay. For example, a typical loam soil is composed of 40% sand, 40% silt, and 20% clay. This yields a balance between aeration vs. drainage vs. water holding capacity vs. fertility.

How a plant works

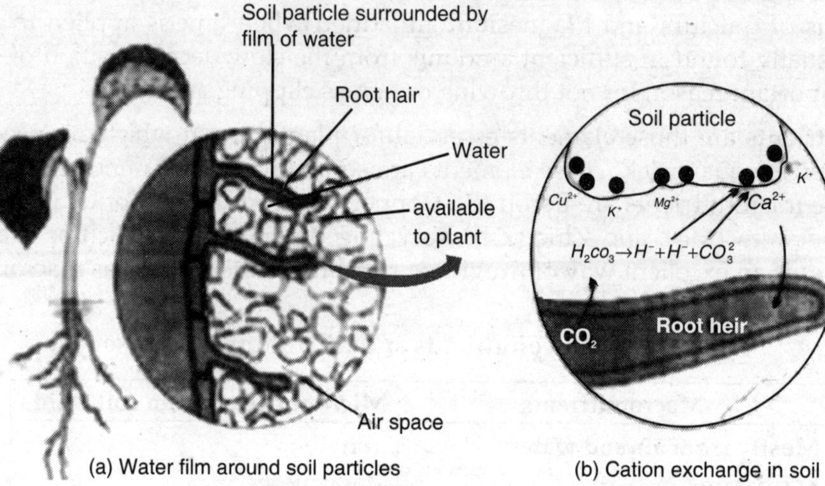

Fig.1.2F$_1$: Working system of a plant

APPENDIX 1.3

PHYSICAL & CHEMICAL PROPERTIES OF SOIL

Physical properties of soil

Physical properties (mechanical behavior) of soil influence its use and behavior towards plant growth. The plant support, root penetration, drainage, aeration, retention of moisture, and plant nutrients are linked with the physical condition of the soil. Physical properties also influence the chemical and biological behavior of soil. The physical properties of a soil depend on the amount, size, shape, arrangement and mineral composition of its particles. These properties also depend on organic matter content and pore spaces. Important physical properties are:

- Texture;
- Particle Density;
- Bulk Density;
- Porosity; and
- Color.

Soil Texture

Soil texture refers to the relative proportion of particle or it is the relative percentage by weight (organic matter is not included) of three soil components-sand, silt, and clay or simply refers to the size of the soil particles. Soils that are preponderantly clay are called clay (textural class); those with high silt content are silt (texture class); those with a high sand percentage are sand (texture class). A soil that does not exhibit the dominant physical properties of any of these three groups (such as a soil with 40 percent sand, 40 percent silt, and 20 percent clay) is called loam. Loam does not contain equal percentages of sand, silt, and clay. It does, however, exhibit approximately equal properties of sand, silt, & clay.

The textural triangle is used to determine the soil texture name such as sandy loams, silty clays, clays, clay loams, etc. The figure shows the soil textural triangle diagram. To use soil textural triangle, start with the sand-size fraction side first, reading the number and line associated with the fraction present, following the link back towards the center of the triangle. Proceed likewise in a clockwise fashion to the clay and silt sides of the triangle, and where all three lines intersect, is a given soil textural classification. For example, a soil containing 30% sand, 30% clay, and 40% silt by weight is classified as a clay loam.

Determining soil texture provides better management guidelines for soil-plant-water relationship, land use capability and methods of soil management, soil stability & readability for tillage guidelines, and root system development & penetration. It influences physical & chemicals properties like water holding capacity, nutrient retention & fixation and its availability, drainage strength, compressibility and thermal regime. Suitability of a particular crop depends on texture in addition to soil depth, depth of water table, salinity. Loamy soils (silty) exhibit intermediate properties and it is best for agriculture production because they retain more water & nutrients than sand soil and have better drainage, aeration & tillage properties than clay soils.

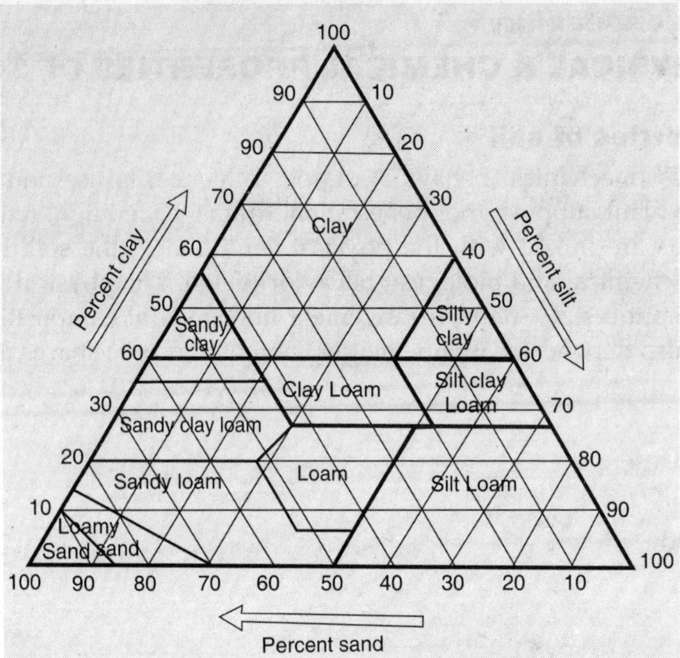

Fig.1.3F_1: Soil textural triangle

Particle Density (PD)

Soil Particle Density (PD) is the weight per unit volume of soil solids. It is the density of the solid soil particles only; the measurement does not include water weight or pore (air) space. PD is not affected by pore space and not related to particle size & arrangement of particles (structure).

Particle densities for most mineral soils = 2.60-2.75 g/cm^3. These soil minerals are-quartz, feldspar, micas, and colloidal silicates.

PD of typical mineral soils= 2.65 mg/m^3.

Bulk Density (BD)

Bulk Density (BD) is the weight of a volume of oven-dry bulk soil (soil Particles + pore space). It is the density for a volume of soil as it exists naturally and includes air space and organic materials in the soil volume. Since bulk density is calculated for the dried soil, moisture is not included in the sample weight. The bulk soil volume is assumed not to have changed by drying; only the water has been removed, leaving empty pores.

BD- changes as the pore space changes;

BD of common surface soils=1.1-1.4 g/cm^3

BD of common sub-soils=1.3-1.7 g/cm^3

Fine-textured soils have lower bulk densities than sandy soils.

Generally, sub-soils have lower bulk densities than surface soils and it affects engineering properties, water movement and rooting depth of plants.

Porosity and Permeability

Pore Spaces (also called voids) in a soil consists of that portion of the soil volume not occupied by solids, either mineral or organic. Pores in soil are the result of irregular shapes of primary particles and their aggregation; the forces of penetrating roots, worms, & insects; and of expanding gases entrapped by water. Under filled conditions, pore spaces are occupied at all times by air and water. Sands have large and continuous pores. In contrast, clays, although containing more total pore space because of the minute size of each clay particle, have very small pores, which transmit water slowly. Porosity is the volume of the pores divided by the bulk volume.

For a given volume of soil; % Pore Space,

$$E_p = 100\% - \left(\frac{\text{bulk density}}{\text{particle density}}\right) \times 00$$

Soil porosity directly influences soil water movement.

Water drains by a gravitational force from pores larger than about 30-60 microns (um). In comparison, root hairs, the smaller than 30 um, attraction forces in the soil retain the water within the fine pores, which result in a waterlogged soil and poor aeration. To the growing plant, pore sizes are more important than total pore space.

Porosity tells us how much water rock or soil can retain. Permeability is a measure of how fast air and water can move through a soil. Both soil texture and structures influence the permeability. Clays reduce permeability because of their small size and flat shape. In addition, movement of clay in the soil can fill the voids clogging the pathways for air and water movement.

Colour

Soil Colour indicates many characteristics of a particular soil. Geologist have recognized over 170 different soil colours. But the most common colour of soils are shades of black, brown, red and grey. A change in soil colour from the adjacent soils indicates a difference in the soil's mineral origin (parent material) or in soil development. White colours are common when salts or carbonated (lime) deposits and warmup more, which aids seed germination.

Color is also an indicator of presence of iron compounds in the soil.

Table 1.3T_1: Major Forms of iron and effect on Soil Colour

Form of Iron	Chemical Formula	Color of soil
Ferrous oxide	FeO	Gray
Ferric oxide (Hematite)	Fe_2O_3	Red
Hydrated ferric oxide	$2Fe_2O_3 . 3H_2O$	Yellow

Table 1.3T_2: Soil Color and their Attributes

Soil Color	Soil Attributes
Brown to black (surface horizon)	Accumulation of OM, humus
Black (subsoil)	Accumulation of Mn Parent material (e.g. basalt)
Yellow to reddish	Fe^{3+} Well-aerated soils or good drainage
Grey, bluish-green	Fe^{2+} Poorly drained soils
White to grey	Accumulation of salts;
White to grey	Parent material: marl, quartz

OM – Organic matter

Chemical Properties of soil

- pH
- Salinity
- Cation matter
- Organic matter
- C: N ratio (carbon to Nitrogen ratio)

Soil pH

Soil pH generally refers to the degree of soil's acidity or alkalinity. Chemically, it is defined as the $\log_{10}$ hydrogen ions (H^+) in the soil solution. The pH scale ranges from 0 to 14; a pH of 7 is considered neutral. If pH values are greater than 7, the solution is considered basic or alkaline; if they are below 7, the solution is acidic. Because the pH scale is in logarithmic units, a change of just a few pH units can induce significant changes in the chemical environment and sensitive biological processes. For example, a soil with pH 5 is 10 or 100 times more acidic than a soil with pH 6 or 7, respectively. Sources of H^+ ions in soil solution includes carbonic acid produced when carbon dioxide (CO_2) from decomposing organic matter, root respiration, and the soil atmosphere is dissolved in the soil water.

Other sources of H^+ ions are root release, reaction of aluminium ions (Al^{+3}) with water, nitrification of ammonium from fertilizers & organic matter mineralization, reaction of sulphur compounds, rainwater, and acid rain.

Soil pH affects the soil's physical, chemical, & biological properties and processes, as well as plant growth. The nutrition, growth, & yields of most crops decrease where pH is low and increase as pH rises to an optimum level(in general, optimal pH is between 5.5-7.5).

Soil Salinity

Soil salinity is the salt content in the soil. In arid region, less rainfall available to leach the salt and high rate of evaporation causes concentration of salt in soil at various layers. Salt can be transported to the soil surface by capillary transport from a salt laden water table and then accumulate due to evaporation. They can also be concentrated in soils due to human activity, for example the use of potassium as fertilizer, which can form sylvite, a naturally occurring salt. As soil salinity increases, salt effects can result in degradation of soils and vegetation. Most crops do not grow well on soils that contain salts. One reason is that salt causes a reduction in the rate and amount of water that the plant root can take up from the soil. Also, some salts are toxic to plants when present in high concentration.

A soil may be rich in salts because the parent rock from which it was formed contains salts. Sea water is another source of salts in low-lying areas along the coast. A very common source of salts in irrigated soils is the irrigation water itself. Most of irrigation water contain some salt. After irrigation, the water added to the soil is used by the crop or evaporates directly from the moist soil. The salt, however, is left behind in the soil. If not removed, it accumulates in the soil; this process is called salinization. Very salty soils are sometimes recognizable by a white layer of dry salt on the soil surface. Salt is a natural element of soils and water. The ions responsible for salinization are: Na^+, K^+, Ca^{2+}, Mg^{2+}, and Cl^+. These salts tend to have very poor structure which limits or prevents water infiltration and drainage. If there is construction of high dam; the change in the level of ground water before the construction had enabled soil erosion, which led to high concentration of salts in the water table. After the construction, the continuous high level of the water table led to the salinization of the arable land.

Salty ground water may also contribute to salinization. When the water table rises the salty ground water may reach the upper soil layers and, thus, supply salts to the root zone.

Soils that contains a harmful amount of salt are referred to as salty or saline soils. Soil, or water, that has a high content of salt is said to have a high salinity. Water salinity is the amount of salt contained in the water. It is also called the "salt concentration" and may be expressed in grams of salt per litre of water (g / l), or in milligrams per litre (which is same as ppm).

However, the salinity of both water and soil is easily measured by means of an electrical device. It is then expressed in terms of electrical conductivity: millimhos / cm or micromhos / cm. A salt concentration of 1 gram per litre is about 1.5 millimhos / cm. Thus a concentration of 3 grams per litre will be about the same as 4.5 millimhos / cm.

The salt concentration in the water extracted from a saturated soil (called saturation extract) defines the salinity of this soil. If this water contains less than 3 grams of salt per litre, the soil is said to be non-saline. If the salt concentration of the saturation extract contains more than 12 g/l, the soil is said to be highly saline.

For improvement of saline soils, more water is applied to the field than is required for crop growth. This additional water infiltrates into the soil and percolates through the root zone. During percolation, it takes up part of the salts in the soil and takes these along to deeper soil layers. In fact, the water washes the salts out of the root zone. This washing process is called leaching. The additional water required for leaching must be removed from the root zone by means of a subsurface drainage system. If not removed, it could cause a rise of the groundwater table which could bring the salts back into the root zone. This, improvement of saline soils includes, essentially, leaching and sub-surface drainage. Thus, salts can be leached by slowly applying excess water.

Soil Electrical Conductivity (EC)

It is a measure of the amount of salts in soil (Salinity of soil). It is an important indicator of soil health. It affects crop yields, crop suitability, plant nutrient availability, and activity of soil microorganisms which influence key soil processes including the emission of greenhouse gases such as nitrogen oxides, methane & carbon dioxide. Excess salts hinder plant growth by affecting the soil-water balance. Soil electrical conductivity is an indirect measurement that correlates very well with several soil physical & chemical properties that affect crop production, including soil texture, cation exchange capacity (CEC), drainage conditions, organic matter level, salinity, & subsoil characteristics. Although EC does not provide a direct measurement of specific ions or salt compounds, it has been correlated to concentrations of nitrates, potassium, sodium, chloride, sulphate, and ammonia. For certain non-saline soils, determining EC can be a convenient and economical way to estimate the amount of nitrogen (N) available for plant growth.

Electrical conductivity is the ability of a material to conduct (transmit) an electrical current and it is commonly expressed in units of milliSiemens per meter (mS/m). Alternatively, electrical conductivity measurements can be expressed in deciSiemens per meter (dS/m), which is 100 times greater than milliSiemens per meter.

EC has not been widely used as a means for measuring or expressing soil salinity for several reasons, Firstly, it is not single-valued; it varies over the irrigation cycle as the soil water content changes. Thus, EC does not lend itself to simple classifications or standards unless it is referenced to specific water content, such as field capacity. Secondly, EC has not been widely adopted for routine appraisals of soil salinity because methods for obtaining soil water samples are not practical at typical field water contents.

The term salinity refers to the presence of the major dissolved inorganic solutes (essentially Na^+, Ca^{++}, K^+, CI, $S04^=$ etc.) in aqueous samples. As applied to salt, it refers to the soluble plus readily dissolvable salts in the soil or, operationally, in an aqueous extract of a soil sample. Salinity is quantified in terms of the total concentration of such soluble salt or more practically, in terms of the EC of the solution, because the two are closely related.

Electrical Conductivity values of different soil materials

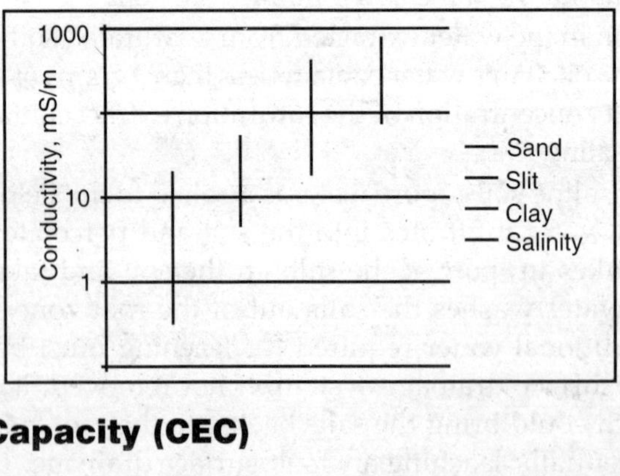

Cation-Exchange Capacity (CEC)

A **cation** is a positively (+) charged ion. Most nutrients are cations: Ca^{2+}, Mg^{2+}, K^+, N H^{4+}, Zn^{2+}, Cu^{2+}, and Mn^{2+}.

Anion is a negatively (-) charged ion.

Cation exchange: A process, in which cation in solution exchange with cations or exchange site of minerals and organic matter (OM).

Cation Exchange Capacity (CEC): The total amount of positive charge that a particular material or soil can potentially absorb at a given pH. These cations are in the soil solution and are in dynamic equilibrium with the cations adsorbed on the surface of clay and organic matter. CEC is a measure of the quantity of cations that can be adsorbed and held by a soil. CEC is dependent upon the amount of organic matter and clay in soils & on the types of clay. In general, the higher OM and clay content, the higher the CEC. It is expressed in terms of positive charge absorbed per unit mass [$cmols_{charge}$ / kg_{soil}] Very high CEC means nutrients holding capacity is high.

Soil Organic Matter (SOM)

When plant residues are returned to the soil, various organic compounds undergo decomposition. Decomposition is a biological process that occurs naturally and includes the physical breakdown and biochemical transformation of complex organic molecules of dead material into simpler organic & inorganic molecules. This can be seen in carbon cycling, which is the continuous transformation of organic & inorganic carbon compounds by plants and micro & macro-organisms between the soil, plants and the atmosphere. In the decomposition process, different products are released: carbon dioxide (CO_2), energy, water, plant nutrients and re-synthesized/less decomposable organic carbon compounds or matters. This release process is called mineralization. The waste products produced by micro-organisms are soil organic matter. This waste material is less decomposable than the original plant and animal material, but it can be used by a large number of organisms.

SOM is the organic fraction of soil exclusive of non-decomposed plant and animal residues. It comprises all of soil components which are derived from plants and animals. Soil Organic Carbon (SOC) is the main source of energy for soil microorganisms. It is simultaneously a source & a sink for nutrients and plays a vital role in soil fertility maintenance.

Thus, soils contain carbon (C) in both organic and inorganic forms. In most soils (with the exception of calcareous soils) the majority of C is held as soil organic carbon (SOC). The term soil organic matter (SOM) is used to describe the organic constituents in the soil. The term 'soil organic carbon' refers to the carbon occurring in the soil in SOM.

C: N Ratio

The ratio of the weight of organic carbon to the weight of total nitrogen in the soil or in organic material, obtained by dividing the percentage of organic carbon (C) by the percentage of total nitrogen (N) is known as C: N ratio. The C: N ratio influences the rate of decomposition of organic matter and is important for practical farming. Organisms that decompose organic matter, use carbon as a source of energy and nitrogen for building cell structure. They need more carbon than nitrogen. Organisms use about 30 parts carbon for each part of nitrogen. If there is too much carbon, decomposition slows when the nitrogen is used up and some organisms die. The course of decomposition of organic matter is affected by the presence of carbon and nitrogen. When the energy source, carbon, is less than that required for converting available nitrogen into protein, organisms make full use of the available carbon and get rid of the excess nitrogen as ammonia. This release of ammonia to the atmosphere *produces* a loss of nitrogen.

The C: N ratio represents the relative proportion of the two elements. A material, for example, having 25 times as much carbon as nitrogen is said to have a C: N ratio of 25:1. Actually the ratio of available carbon to available nitrogen is the important relationship because there may be some carbon present which so resistant to biological attack that its presence is not significant.

APPENDIX 1.4

SOIL EROSION: TYPES, CAUSES & PREVENTION

The loss of nutrient rich top soil or disturbances to the soil structure by some natural processes or anthropogenic activities is known as soil erosion. Soil erosion is a part of the soil pollution. The soil erosion is creeping death of the soil. When the rate of removal of soil exceeds the rate of soil formation, the soil erosion occurs.

The soil erosions are of two types

(i) Natural or geological or background soil erosion

(ii) Accelerated soil erosion

The natural soil erosion takes place by normal or natural conditions without any interference of man. But the accelerated soil erosion is due to active interference of man and animals in disturbing the equilibrium in between the rate of soil formation and rate of soil loss. The magnitude of soil erosion is governed by the factors like corrosivity of the eroding agent, erodibility of the soil, slope, steepness & length, land use practices and conservation strategies employed.

Causes of soil erosion

The process of soil erosion is caused by the following agents:

(a) Water erosion: Water is powerful agent causing soil erosion either in the form of rain drops or in the form of surface flow. When the soil is lost like a thin covering from a large surface area, the process is known as sheet erosion. When the sheet erosion takes place with full momentum, leading to cutting of soil surface as thin channels, it is known as rill erosion. Several rills from the steep slope converge forming wider channel known as gully erosion.

(b) Wind erosion: In dry (arid) regions, the soil is mainly sandy and the vegetation is grossly inadequate. In such places, wind causes loss of soil.

(c) Land slide or slip erosion: When the rocks at cliffs face heavy rains, their weights undergo sharp rise due to gravitational forces. Such rocks with high weights slip over lower rocks causing land slide and there by slip erosion.

(d) Stream bank erosion: During flood, the rivers splash their water against banks and cut through them taking away soil from the bank.

(e) Deforestation and overgrazing: Loss of natural vegetation due to deforestation and overgrazing leads to dry & dust soil which can be taken away easily by wind and water.

(f) Mining erosion: A lot of soil is also eroded due to mining and industrial activities. Since soil is very precious for the existence of living organisms, it is the need of the hour to prevent the loss of soil by adopting suitable mechanisms.

The process of soil loss can be checked by the following means

- Protection of soil from the severity of rain drops through afforestation.
- Preventing water from concentrating and passing through the slopes.

- Slowing down the momentum of water flow adopting various scientific techniques.
- Growing vegetation covers to minimize wind velocity.
- Growing soil binders like grasses to prevent direct contact in between soil and eroding agents.

The contribution of different global agent towards total global soil degradation (erosion) are as follows:

- Overgrazing accounts for 34.5% of the degraded area.
- Deforestation accounts for 29.5%.
- Agricultural activities accounts for 28.1 %.
- Over exploitation of soil accounts for 7.0%.
- Bio-industrial activities accounts for 1.2%.

Soil erosion is always a result of mankind's unwise actions, such as overgrazing or unsuitable cultivation practices. These leave the land unprotected and vulnerable. Accelerated soil erosion by water or wind may affect both agricultural areas and the natural environment. Soil erosion is just one form of soil degradation. Other kinds of soil degradation include salinization, nutrient loss, and compaction.

Prevention of soil erosion: Plants provide protective cover on the land and prevent soil erosion for the following reasons

(a) Plants slow down water as it flows over the land (runoff) and this allows much of the rain to soak into the ground;
(b) Plants roots hold the soil in position and prevent it from being washed away;
(c) Plants break the impact of a raindrop before it hits the soil, thus reducing its ability to erode;
(d) Plants in wetlands & on the banks of rivers are of particular importance as they slow down the flow of the water and their roots bind the soil, thus preventing erosion.

Preventing soil erosion requires technical changes to adopt. Aspects of technical changes include:

(i) use of contour ploughing and wind breaks;
(ii) leaving unploughed grass strips between ploughed land;
(iii) making sure that there are always plants growing on the soil, and that the soil is rich in humus (decaying plant and animal remains). This organic matter is the "glue" that binds the soil particles together and plays an important part in preventing erosion;
(iv) avoiding overgrazing and the over-use of crop lands;
(v) allowing indigenous plants to grow along the river banks instead of ploughing and planting crops right up to the water's edge;
(vi) encouraging biological diversity by planting several different types of plants together;
(vii) conservation of wetlands.

Soil erosion can be checked by adopting the following additional practices:

- Intensive cropping and use of proper drainage canals.
- Terracing on the sloping fields. This retards the speed of the flowing water.
- Planting trees and sowing grasses.
- Extensive afforestation practices to be carried out.

2 CHAPTER Introduction to Soil Pollution

Land/Soil pollution means the buildup in soils of persistent toxic compounds, chemicals, salts, radioactive materials, or disease causing agents, which have adverse effects on plant growth and animal health. Soil pollution is the degradation of earth's land surface caused by human activities and their misuse of land resources. Urbanization and industrialization are major causes of soil pollution.

Soil is the thin layer of organic and inorganic material that covers the earth's rocky surface. The organic portion which is derived from the decayed remains of plant and animals, is concentrated in the dark uppermost top soil. The inorganic portion made up of rock fragments, was formed over thousands of years by physical and chemical weathering of bedrock. Productive soil is necessary for agriculture to supply the world with sufficient food.

Different ways by which soil can become polluted are:

- Improper disposal of waste and exploitation of minerals
- Seepage from a landfill
- Discharge of industrial waste into the soil
- Percolation of contaminated water into the soil
- Rupture of underground storage tanks
- Excess application of pesticides, herbicides and fertilizers
- Solid waste seepage
- Improper use of soil by inadequate agriculture practices

The most common chemicals involved in causing soil pollution are:

- Petroleum hydrocarbons
- Heavy metals
- Pesticides
- Solvent

Types of Soil Pollution

(a) Agricultural Soil Pollution
  (i) Pollution of surface soil
  (ii) Pollution of underground soil

(b) Soil pollution by industrial effluents and solid wastes:
  (i) Pollution of surface soil
  (ii) Disturbance in soil profile

(c) Pollution due to urban activities:
   (i) Pollution of surface soil
   (ii) Pollution of underground soil

Causes of soil Pollution

Soil pollution is caused by the presence of man-made chemicals or other alteration in the natural soil environment. This type of contamination typically arises from the rupture of underground storage tanks, application of pesticides, pesticides percolation of contaminated surface strata, oil tanks and fuel dumping, leaching of wastes from landfills or direct discharge of industrial wastes to the soil. The most common chemicals involved are petroleum hydrocarbons, solvents, pesticides, lead and other heavy metals. The Occurrence of this phenomenon is correlated with the degree of industrialization and intensities of chemicals usage.

A soil pollutant is any factor which deteriorates the quality, texture and mineral content of the soil or which disturbs the biological balance of the organisms in the soil. Pollution in soil has adverse effect on plant growth.

Pollution in soil is associated with:

(a) Indiscriminate use of fertilizers.
(b) Indiscriminate use of pesticides, insecticides and herbicides.
(c) Dumping of large quantities of solid waste.
(d) Deforestation and soil erosion.
(e) Pollution due to industrial activities.
(f) Pollution due to Urbanization

(a) Indiscriminate use of fertilizers

Soil nutrient are important for plant growth and development. Plants obtain carbon, hydrogen & oxygen from air and water. But other necessary nutrients like nitrogen, phosphorus, potassium, calcium, magnesium, sulphur and more must be obtained from the soil. Farmers generally use fertilizers to correct soil deficiencies. Fertilizers contaminate the soil with impurities, which come from the new raw materials used for their manufacture. Mixed fertilizers often contain ammonium nitrate (NH_4NO_3), phosphorus as P_2O_5, and potassium as K_2O. For instance, As , Pb and Cd present in trace in rock phosphate mineral get transferred to super phosphate fertilizer. Since the metals are not degradable, their toxic levels due to excessive use of phosphatic fertilizers, becomes an indestructible poison for crops.

The over use of fertilizers reduce quantity of vegetables and crops grown down on soil over the years. It also reduces the protein content of wheat, maize, gram, etc., grown on that soil. The carbohydrate quality of such crops also get degraded. Excess potassium content in soil decreases Vitamin C and carotene content in vegetables & fruits. The vegetables and fruits grown on over-fertilized soil are more prone to attack by insects and disease.

(b) Indiscriminate use of pesticides, insecticides and herbicides

Plants which depend on food are under attack from insects, fungi, bacteria, viruses, rodent &other animals, and must compete with weed for nutrients. To kill unwanted population living in crops, farmer use pesticides. The first widespread insecticide use began at the end of World War II and included DDT (dichloro diphenyl trichloroethane) and gammaxene. Insects soon became resistant to DDT and as the chemical did not decompose readily, it persisted in the environment. Since it was soluble in fat rather than water, it biomagnified up the food chain and disrupted calcium metabolism in bird, causing eggshell to be thin & fragile. As a result, large birds of prey such as brown pelican, osprey, falcons and eagles became endangered. DDT has been now been banned in most countries. Ironically many of them including USA, still produce DDT for export to other developing nations whose need outweigh the problems caused by it.

The most important pesticides are DDT, BHC, chlorinated hydrocarbons, organophosphates, aldrin, malathion, dieldrin, etc. The remnants of such pesticides used on pests may get adsorbed by the soil particles, which then contaminates root crops grown in that soil. The consumption of such crops causes the pesticide remnants to enter human biological systems, affecting them adversely.

An infamous herbicides used as defoliant in the Vietnam War called Agent Orange (dioxin), was eventually banned. Soldier's cancer cases, skin condition and their biodegradation may take weeks & even months.

Pesticides problems such as resistance, resurgence, and health effects have caused scientists to seek alternatives. Pheromones and hormones to attract or repel insects and using natural enemies or sterilization by radiation have been suggested.

(c) Dumping of Solids Wastes

Solid waste includes garbage, domestic refuse and discarded solid material such as those from commercial, industrial & agriculture operations. They contain increasing amounts of paper, cardboards, plastics, glass, old construction materials, packing material and toxic or otherwise hazardous substance. Since a significant amount of urban solid waste tends to be paper and food waste, the majority is recyclable or biodegradables in landfills. Similarly, most agricultural waste is recycled and mining waste is left on site.

The portion of solid waste that is hazardous such as oils, battery metals, heavy metals from smelting industries and organic solvents are the ones we have to pay particular attention to. These can in long run, get deposited to the soil of the surrounding area and pollute them by altering their chemical & biological properties. They also contaminate drinking water aquifer sources. More than 90% of hazardous waste is produced by chemicals, petroleum & metal-related industries and small businesses such as dry cleaners & gas station contributes as well.

(d) Deforestation and Soil Erosion

Soil Erosion occurs when the weathered solid particles are dislodged and carried away by wind or water.

Deforestation, agricultural development, temperature extremes, precipitation including acid rain, and human activities contribute to this erosion. Humans speed up this process by construction, mining, cutting of timber, over cropping and overgrazing. It results in floods and cause soil erosion.

Forests and grassland are excellent binding materials that keeps the soil intact & healthy. They support many habitats and ecosystems, which provide innumerable feeding pathways or food chains to all species. Their loss would threaten food chains and the survival of many species. During the past few years, lot of vast green land has been converted into deserts. The precious rain forest habitats of South American, tropical Asia and Africa are coming under pressure of population growth & development (especially timber, construction and agriculture) as many believe that a wealth of medicinal substance including a cure for cancer & aids, lie in these forest. Deforestation is slowly destroying the most productive flora and fauna areas in the world, which also from vast of a very valuable sink for CO_2.

(e) Pollution due to Industrial Activities

Industrial activities also are contributing factor to land pollution. Industrial wastes are the effluent discharged from chemical industries, paper & pulp mills, tanneries, textile mills, steel industries, distilleries, refineries, pesticides & fertilizers industries, pharmaceutical industries, food processing industries, cement industries, thermal & nuclear power plants, mining industries etc.

Thermal power plants generate a large quantity of 'Fly ash'. Huge quantities of these wastes are dumped on land which cause land pollution. Mining also affects ground & surface waters, the aquatic life, vegetation, soils, animals and the human health. Acid mine drainage can cause damage to streams which in return can kill aquatic life. Mining gas and petroleum also pollute the land. Petroleum extraction and manufacturing contaminates the soil with bitumen, gasoline, kerosene & mining brine solutions. Open cast mining, which is a process where the surface of the earth is dug open to bring out the underground mineral deposits, destroys the topsoil and contaminates the area with toxic metals & chemicals.

(f) Pollution due to Urbanization

Urban activities generate large quantities of city wastes including several biodegradable materials (like vegetables, animal's wastes, papers, wooden pieces, caresses, plant twigs, leaves, cloth wastes as well as sweepings) and many non-biodegradable materials (such as plastic bags, plastic bottles, plastic wastes, glass bottles, glass pieces , stone/cement pieces). On a rough estimate Indian cities are producing solid city wastes to the tune of 80,000 metric tons every day. If left uncollected and undecomposed, they are a cause of several problems such as:

- **Clogging of drains:** Causing serious drainage problems including the burst/leakage of drainage lines leading to health problems.
- **Barrier to movement of water:** Solid wastes have seriously damaged the normal movement of water thus creating problem of inundation, damage to foundation of buildings as well as public health hazards.
- **Foul smell:** Generated by dumping the wastes at a place.
- **Increased microbial activities**: Microbial decomposition of organic wastes generate large quantities of methane besides many chemicals to pollute the soil and water flowing on its surface.
- When such solid wastes are hospital wastes they create many health problems as they may have dangerous pathogen within them besides dangerous medicines, injections etc.

Pollution of Underground Soil

Underground soil in cities is polluted by:

- Chemicals released by industrial wastes.
- Decomposed and partially decomposed materials of sanitary wastes.

Many dangerous chemicals like cadmium, chromium, lead, arsenic, selenium products are likely to be deposited in underground soil. Similarly, underground soil polluted by sanitary wastes generate many harmful chemicals. These can damage the normal activities and ecological balance in the underground soil.

Causes of soil pollution can be summarized as follows:

- Polluted water discharged from factories.
- Runoff from pollutants (paints, chemicals, rotting organic material) leaching out of landfill.
- Oil and petroleum leaks from vehicles washed off the road by the rain into the surrounding habitat.
- Chemical fertilizer runoff from farms and crops.
- Acid rain (fumes from factories mixing with rain).
- Sewage discharged into rivers instead of being treated properly.
- Over application of pesticides and fertilizers.
- Purposeful injection into groundwater as a disposal method.
- Interconnections between aquifers during drilling (poor technique).
- Septic tank seepage.
- Sanitary hazardous landfill seepage.
- Lagoon seepage.
- Cemeteries.
- Scrap yards (waste oil and chemical drainage).
- Leaks from sanitary sewers.

Soil pollutants

(a) Pesticide pollution

In modern agriculture the use of various agrochemicals is a common practice. These include pesticides, herbicides, insecticides, fungicides and others. Pesticides applied on seed or foliage ultimately reach the soil. Accumulation of pesticide residues in the biosphere creates ecological stress causing contamination of soil, water, and food. Persisting chemicals may also be hazardous to human health and should be eliminated. Persistent pesticides may accumulate in the bodies of animals and over a period of time increase in concentration if the animal is unable to flush leading to bioaccumulation. When an affected animal is eaten by a carnivore, the pesticide is further concentrated in the carnivore. This phenomenon ie. Increase in the concentration of a nondegradable substance along the food chain is called Bio magnification.

Another problem associated with insecticides is the ability of insects to become resistant. Most pesticides kill beneficial predators and parasites. The short term & long term health effects to the persons using the pesticides and public that consumes the food are the major concerns. Exposure to small quantities for longer time causes mutations leading to cancer. Pesticides or their metabolites affect many soil microbes and their activities. Seed treatment with mercuric fungicides are found to be inhibitory to Rhizobium (nodulation and nitrogen fixation), Nitrosamines and Nitrobacterium (nitrification).

(b) Fertilizer pollution

The agricultural production depends on chemical fertilizer application, as most of our high yielding varieties are fertilizer responsive. Continuous application of chemical fertilizers alone lead to deterioration of soil properties and cultivated soils loose their natural characteristics. Fertilizers like ammonium sulphate, ammonium chloride and urea reduce the soil pH. Many crops, like potato, grapes, citrus, beans are sensitive to chloride toxicity. In integrated nutrient management, to sustain the productivity of our soils, organic manures and bio fertilizers are recommended as supplements to chemical fertilizers.

(c) Nitrate pollution

Nitrogen occurs in many forms in the environment and takes part in many biochemical reactions. The four forms of nitrogen that are of particular significance in environmental technology are organic nitrogen, ammonia nitrogen, nitrite nitrogen, and nitrate nitrogen. In water contaminated with sewage, most of the nitrogen is originally present in the form of complex organic molecules (protein) and ammonia (NH_3). These substances are eventually broken down by microbes to form nitrites and nitrates.

Nitrogen, particularly in the nitrate form, is a basic nutrient that is essential to the growth of plants. Excessive nitrate concentrations in surface waters encourage the rapid growth of microscopic plants called algae and excessive growth of algae degrades water quality.

Nitrates can enter the ground water from chemical fertilizers used in agricultural areas. Excessive nitrate concentrations in drinking water pose an immediate and serious health threat to infants under 3 months of age. The nitrate ions react with blood hemoglobin, reducing the blood's ability to carry oxygen and this produces a disease called ***blue baby*** or *methemoglobinemia.*

- An illness that arises when an infant's blood is unable to carry enough oxygen to body cells and tissue.
- An infant with moderate to serious "blue baby syndrome" may have a brownish-blue skin due to lack of oxygen.
- Child may be fussy, tired, have diarrhea or vomiting.
- Severe cases can cause death.

Low oxygen levels in the blood cause the lips, fingers, and toes to took blue (cyanotic)

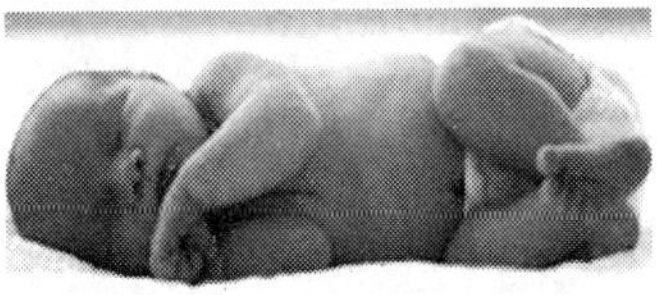

Fig 2.1: Blue baby syndrome

(d) Excess Salts and Water

Irrigation water helps to produce more yield than rain fed land. Irrigation water contains dissolved salts and in dry season, water is in the form of saline solution evaporates leaving its salts such as NaCI in the top soil. This saline soil causes stunted plant growth, lower yield. Flushing out salts reduces the salinity but makes downstream irrigation water, saltier. Another problem is water logging.

(e) Heavy metal pollution

Heavy metals include all metals with atomic numbers greater than 23 (with few exceptions) or density more than 5 gm per ml. Heavy metals are hazardous, not acceptable to biological system. They are toxic to man and other life forms. Most of them are slow poisons as they accumulate in the body and cause serious disorders. Mercury, lead, arsenic, chromium & cadmium are the five most common toxic heavy metals and they have serious effects on human health.

Table 2.1: Sources and effects of heavy metals pollution

S.No	Heavy metals(forms)	Sources	Effects
1.	Mercury: Hg^{++} (Mercuric) $C_6H_5Hg\ CH_3\ COO$	Methyl mercury fungicides, electrical and electronic industries, PVC, Plastics, Paints.	Irreversible neurological damage in man, Minamata disease
2.	Lead Pb^{2+}, Pb^{4+}	Automobile exhaust of leaded petrol (50%), Batteries, Pipes, Soldiering.	Cause mutation in algae and bacteria, blackening in fish, gradual paralysis in man.
3.	Arsenic As^{+++} Arsenic trioxide, Sodium arsenate	Herbicides, fungicides, wood preservative – Agro chemicals (70%), industrial chemicals – paints, bullets (20%), glass and wares (5%).	Accumulated in hair, nail, skin lesions, act as oxidative uncoupler, cause damage to kidney, respiratory tract & nervous disorders.
4.	Chromium Cr^{+6} CrO_3	Tanneries, electroplating and metal finishing processes, khaki dyeing in textiles.	Toxic to aquatic organisms, absorbed through intestinal tract in man.
5.	Cadmium (Cd)	Pigment and stabilizer for PVC, plastics, tires, rechargeable cells, electroplating, coal, oil and phosphate rocks.	Bones become brittle – Itai itai disease in japan, gastro enteric distress and pain.

The unique physical, chemical & toxic properties of heavy metals have promoted their wide use in industrial processes and as biocides (fungicide and herbicide). As a result, higher concentration of these heavy metals accumulates in the environment, causing public health hazards and ecological problems. Removal of these metals is therefore a challenge to environmental management. The metals are generally removed by ion exchange & sorption to resins and precipitation as metal sulphides. Biodegradation of metals is not possible, because unlike organic pollutants, metals as elements cannot be mineralized to non-toxic compounds such as H_2O and CO_2. However, bio mobilization is a valid concept in the management of metal pollution. Eukaryotic organisms detoxify heavy metals by binding to polythiols and bacteria have developed different and efficient mechanisms for tolerating heavy metals. They carry the genes controlling metal resistance on chromosome and plasmids. Plasmids often contain genes resistance to several metals (Hg, Pb, As, Cr, Cd, Mo, U). As a result of biological action, metals undergo changes in valency and or conversion into organo metallic compounds.

(f) Industrial Wastes

Indiscriminate dumping of untreated or inadequately treated domestic, mining & industrial wastes on land is an important source of soil pollution. Fall out of gaseous and particulate air pollutants from mining & smelting operations, smoke stacks etc. are the major source of soil pollutants in nearby areas.

(g) Urban Wastes

Millions of tons of urban wastes are produced every year from polluted cities. The inadequately treated or untreated sewage sludge not only poses serious health hazards but also pollutes soil and decreases its fertility & productivity. Other waste materials such as rubbish, used plastic bag, garbage sludge, dead animals, hospital wastes, skins, tyres , shoes etc. cause land and soil pollution. Suspended matter present in sewage can act as a blanket on the soil and interfere with its productivity.

(h) Plastics

Plastic forms a major part of the global domestic & industrial waste. Being non-biodegradable, waste plastic accumulates, adding to pollution. Using photodegradable plastic or biodegradable plastic can solve plastic pollution problem. Photodegradable plastic contains an element which is sensitive to UV rays. Under The effect of solar rays the element is activated and breaks the polymeric chain of the photodegradable plastic. It results in small fragments that are easily digested by microbes.

Various sources of land/soil pollution are tabulated below.

Table 2.2: Sources of land/soil pollution

S. No.	Sources of Land Pollution	Methods of land pollution
1.	Agricultural sources	accumulation of animals manures. excessive input of chemical fertilizers. illicit dumping of tainted crops on land
2.	Mining and Quarrying	use of explosives to blow up mines, use of machineries which emits toxic by products and leaks to the ground.
3.	Sewage sludge	improper method of dredging at fertile land causes soil infertility, leaving the soil more prone to external pollution.
4.	Dredged spoils	improper method of dredging at fertile land causes soil infertility, leaving the soil more prone to external pollution.
5.	Household	improper waste disposal system causes waste accumulation. Improper sanitation system.

6.	Demolition and construction	Non-biodegradable rubbles/debris which are not cleared settle in the soil.
7.	Industrial sources	Poisonous/toxic gaseous emission undergo chemical reactions and increase soil toxicity.

Effects of Soil Pollution

(a) Agriculture

- Reduced soil fertility.
- Reduced nitrogen fixation.
- Increased erodibility.
- Larger loss of soil and nutrients.
- Deposition of silt in tanks and reservoirs.
- Reduced crop yield.
- Imbalance in soil fauna and flora.
- Destruction of vegetable that provides food and shelter

(b) Industrial

- Dangerous chemicals entering underground water.
- Ecological imbalance.
- Release of pollution gases.
- Release of radioactive rays causing health problems.
- Increased salinity.
- Reduced vegetation.

(c) Urban

- Clogging of drains.
- Public health problems.
- Inundation of areas.
- Public health problems.
- Pollution of drinking water sources.
- Foul smell and release of gases.
- Waste management problems.

Long Term Effects of Soil Pollution on Environment

Soil that has been contaminated should not be used to grow food, because the chemicals can leach into the food and harm people who eat it.

If contaminated soil is used to grow food, the land will usually produce lower yield than it would, if it were not contaminated. This, in turn, can cause even more harm because a lack of plants on the soil will cause more erosion, spreading the contaminants onto land that might not have been polluted before.

In addition, the pollution will change the makeup of the soil and the types of microorganisms that will live in it. If certain organisms die off in the area, the larger predator animals will also have to move away or die because they have lost their food supply. Thus it is possible for soil pollution to change whole ecosystems.

Effects of soil pollution are summarized below:

- Pollution runs off into rivers and kills fish, plants & other aquatic life.
- Crops and fodder grown on polluted soil may pass the pollutions on the consumers.
- Polluted soil may no longer grow crops and fodder.
- Soil structure is damaged (clay ionic structure impaired).
- Corrosion of foundations and pipelines.
- Impairs soil stability.
- May release vapours and hydrocarbon into buildings & cellars.
- May create toxic dust.
- May poison children playing in the area.

Pesticides can damage crops; kill vegetation; and poison birds, animals, & fish. Most pesticides kill or damage life from other than those intended. For example, pesticides used in an effort to control or destroy undesirable vegetation and insects often destroy birds & small animals. Some life forms develop immunity to pesticides used to destroy them.

- Acid rain kills trees and other plants.
- Exterminates wild life.

Control of Soil Pollution

The following steps have been suggested to control soil pollution. To help prevent soil erosion, we can limit construction in sensitive area. We will need less fertilizers and pesticides, if we would adopt the three Rs: Reduce, Reuse, and Recycle. This would give us less solid waste.

(a) Reducing chemical fertilizers and pesticide use

Applying bio-fertilizers and manures can reduce chemical fertilizers & pesticide use. Biological methods of pest control can also reduce the use of pesticides and thereby minimize soil pollution.

(b) Reusing of materials

Materials such as glass containers, plastic bags, paper, cloth etc. can be reused at domestic levels rather than being disposed, reducing solid waste pollution.

(c) Recycling and recovery of minerals:

This is a reasonable solution of reducing soil pollution. Material such as paper, some kinds of plastic and glass can be recycled. This decreases the volume of refuse and helps in the conservation of natural resources. For example, recovery of one tonne of paper can save 17 trees.

(d) Reforesting

Control of land loss and soil erosion can be attempted through restoring forest & grass cover to check wastelands, soil erosion and floods. Crop rotation or mixed cropping can improve the fertility of the land.

(b) Solid waste treatment

Proper methods should be adopted for management of solid waste disposal. Industries wastes can be treated physically, chemically and biologically until they are less hazardous. Acidic and alkaline wastes should be first neutralized; the insoluble material if biodegradable should be allowed to degrade under controlled conditions before being disposed.

As a last resort, new areas for storage of hazardous waste should be investigated such as deep well injection and more secure landfills. Burying the waste in locations situated away from residential areas is the simple and most widely used technique of solid waste management. Environmental and aesthetic considerations must be taken into consideration before selecting the dumping sites.

Incineration of other wastes is expensive and leaves a huge residue that adds to air pollution. Pyrolysis is a process of combustion in absence of oxygen or the material burnt under controlled atmosphere of oxygen. It is an alternative to incineration. The gas and liquid thus obtained can be used as fuels. Pyrolysis of carbonaceous waste like firewood, coconut, palm waste, corn combs, cashew shell, rice husk, paddy straw & saw dust, yields charcoal along with products like tar, methyl alcohol, acetic acid, acetone and a fuel gas.

Anaerobic & aerobic decomposition of biodegradable municipal & domestic waste is also being done and gives organic manure. Cow dung which release methane into the atmosphere, should be processed further in 'gobar gas plants' to produce gobar gas' and good manure.

Natural Land Pollution

Land pollution occurs massively during earth quakes, land slides, hurricanes and floods. All cause hard to clean mess, which is expensive to clean, and may sometime take years to restore the affected area. These kinds of natural disaster are not only a problem in that they cause pollutions but also because they render many victims homeless.

APPENDIX 2.1

Major routes of Land Pollution

S. No.	Route of land pollution	Major Pollutants	Remarks
1.	Application of pesticides to agriculture land and forests	Insecticides, rodenticides, herbicides, fungicides as sprays, dusts, seed dressing etc.	There are strict regulations for controlling the application of pesticides
2.	Control of insect vectors of disease	Insecticides	Major pollution over large areas as a consequence of control measures against malarial mosquito & fly.
3.	Application of sewage to agricultural land.	Heavy metals, nitrates detergent	–
4.	Flooding by rivers or seas	A variety of pollutants including those associated with sewage	–
5.	Waste dumping including rubbish dumps/landfill sites/industrial dumps	A very wide range of different pollutants	Some industrial dumps are high in particular pollutants e.g., oil, metal deposits, PCB's.
6.	Precipitation from air as dusts or droplets or in rain or snow	Pollutions associated with soot & dusts, acid rain, pesticides	Transport of pollutants may be over short distances (spray drift, soot and dust from chimney) or long distance (brought down especially by rain and snow)

Note: Of all our resources, land is the most tangible one. Land, which is the solid part of earth's surface is a finite resource, so great care should be taken to preserve it. Land has been put to many new uses, apart from traditional ones.

These are the general uses of land:

- Agriculture and horticulture
- Housing
- Laying of roads and railway lines
- Industries

- Mining of mineral resources
- A container for holding water
- For fodder, cattle grazing, forest wealth for timber, fuel, etc.

The above list gives an overview of India's land classification. The total area of the country is 329 million hectares (mha). Of these 266 million hectares are available for potential use. But in reality, the land has been subjected to varying degrees of degradation. Out of 266 mha available for potential use, 175 mha are degraded due to one or more of the reasons mentioned below:

- Land suffering from serious water and wind erosion-150mha.
- Land affected by salinity & alkalinity – 7mha.
- Land affected by water logging -6mha.
- Land affected by shifting cultivation – 3mha.
- Land affected by miscellaneous factors – 9mha.

Most human or natural activities need space for their location and development, which is provided by land. The various purpose for which land can be used, includes agricultural, human dwelling, industrial purpose, waste disposal, forest, etc.

The land use pattern of India's is as Under:

- Agricultural Land: 43.6%
- Permanent Pastures and meadows: 4.6%
- Cultural wastelands: 12.2%
- Forests: 10.7%
- Barren and uncultivable land: 8.4%
- Urban Land: 5.3%
- No information available: 5.2%

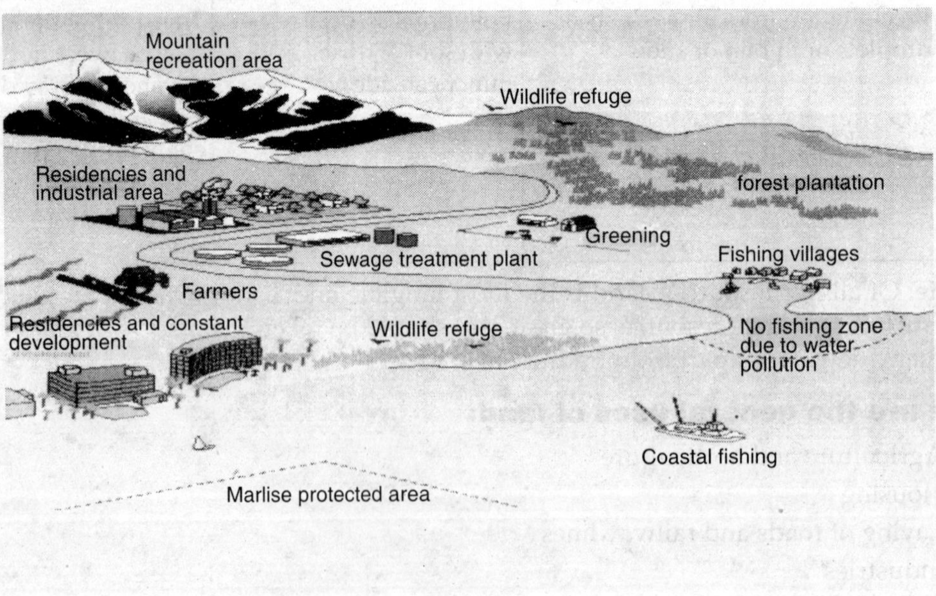

Causes of Land Degradation

Soil degradation is the loss of top soil which erodes its fertility & reduces its water holding capacity. Land Degradation is the temporary or permanent lowering of the productive capacity of the land.

Major causes of land degradation are:

(a) **Soil erosion:** Soil erosion means wearing away of soil. It refers to loss of the superficial layer of the soil by the action of wind, water or human activities.
- The top layer of soil is very fertile. Therefore, its removal degrades the quality of land making it less productive.
- Soil erosion is caused due to heavy rainfall, lack of vegetation cover, and soil mismanagement like, uncontrolled grazing by cattle, faulty method of surface drainage, removal of forest cover etc.

(b) **Salination:** Salination refers to increase in the concentration of soluble salts in the soil due to poor irrigation practices . Flood water results in accumulation of dissolved salts on the soil surface. In dry areas water is evaporated quickly leaving behind a white crust of salts in soil on the surface. The higher concentration of salts severely affects the water absorption process of the plants, resulting into poor production.

(c) **Water Logging:** Water logging may be due to surface flooding or due to high water table. Excessive irrigation particles may cause water logging due to rise in water table of the area. The productivity of water logged soil is severely affected due to lesser availability of plants.

(d) **Shifting Cultivation:** Among many tribal communities, there is a very peculiar practice of cutting down a selected portion of the forest and burning. Crops were cultivated for a few years till the fertility of soil is reduced. When the fields become sterile they were abandoned & a fresh patch of forest were burnt and fresh fields were formed for crop cultivation.

(e) **Urbanization:** Man is to be blamed for most of the land degradation. Productive areas are fast reducing because of developmental activities such as human settlement, industries, roads, railways, airports etc.

(f) **Dumping:** Dumping of solid wastes is one of the most important factors which are responsible for the land pollution. In developing countries like India, the garbage and refuse products are simply dumped leading to land pollution.

(g) **Other causes of land degradation include:** excessive farming, construction, overgrazing, burning of grass cover, deforestation and excessive use of fertilizers & pesticides.

ENVIRONMENTAL TRENDS AND RESPONSES

Land-use changes have had both positive & negative effects on human well-being, and on the provision of ecosystem services. The enormous increase in the production of farm & forest products has brought greater wealth and more secure livelihoods for billons, but often at the cost of land degradation, biodiversity loss & disruption of biophysical cycles, such as the water & nutrient cycles. These impacts create many challenges and opportunities. Table below summarizes positive & negative links between changes in land and human well-being.

Table 2.1T$_1$: Positive & negative links between changes in land and human well-being

S.N	Change in land	Environmental impact	Material needs	Human health	Safety	Socio-economic
1	Cropland expansion and intensification	Loss of habitat and biodiversity; soil water retention and regulation; disturbance of biological cycle; increase of soil erosion, nutrient depletion, salinity, and eutrophication	Increased food and fiber production such as doubling world grain harvest in last 40 years Competing demands for water	Spread of disease vectors related to vegetation and water (such as irrigation associated with schistosomiasis) Exposure to agrochemicals in air, soil and water	Increased hazards from dust and land-slides during extreme weather	More secure livelihoods and growth in agricultural output Changes in social and power structures
2	Loss of forest grassland and wetlands	Loss of habitat, biodiversity, stored carbon, soil water retention and regulation Disturbance of biological cycles and food webs	Diminished variety of resources and water quality	Loss of forest ecosystem services, including potential new medicinal products	Increased hazard of flooding and land-slides during extreme weather and tsunamis	Loss of forest products, grazing fisheries and drought reserves Loss of livelihood, cultural values and support for traditional lifestyles of indigenous & local communities Loss of recreation opportunities and tourism
3	Urban expansion	Disruption of hydrological and biological cycles; loss of habitat and biodiversity;	Increased access to flood, water and shelter; increased choice, but	Respiratory and digestive-tract diseases due to air pollution, poor water supply and sanitation	Increased exposure to crime Traffic and transport hazards	Increased opportunity for social & economic interaction and access to services

			concentration of pollutants, solid and organic wastes; urban heat islands	satisfaction of material needs highly depended on income	Higher incidence of stress and industry-related diseases, Higher incidence of heat stroke	Increased hazard of flooding caused by soil sealing and occupation of hazardous sites	Increased competition for financial resources, Diminished sense of isolation
4	4a	Chemical contamination	Polluted soils and water	–	–	Increased risk of exposure and of contamination of food chains; in severe cases, areas become uninhabitable	Loss of productivity due to ill health Diminished productivity of contaminated systems
	4b	Soil erosion	Loss of soil, nutrients, habitat and, property; siltation of reservoirs	Loss of food and water security	Hunger, malnutrition, exposure to diseases due to weakened immune system Turbidity and contaminated water	Risk of floods and land-slides Accident due to damage to infrastructure particularly in coastal and riverine areas	Loss of property and infrastructure Decreasing hydro power generation due to siltation of reservoirs Diminished development in farm and forest sectors
	4c	Nutrient depletion	Impoverished soils	Diminished farm and forest production	Malnutrition and hunger	–	Lack of development poverty
	4d	Water scarcity	Diminished stream flow and groundwater recharge	Loss of food water security	Dehydration Inadequate hygiene. water related disease	Conflict over water resources	Lack of development, poverty
	4e	Salinity	Unproductive soils, unusable water resources, loss of fresh water habitat	Diminished farm production	Non-portable water	–	Loss of production, Increased industrial costs of corrosions and water treatment, Damage to infrastructure

5	Desertification	Loss of habitat and biodiversity Reduced groundwater recharge, water quality and soil fertility Increased soil erosion dust storms, and sand encroachment	Diminished farm and rangeland production Loss of biodiversity Water scarcity	Malnutrition and hunger Water-borne disease, respiratory problems	Conflict over land and water resources Increasing flash floods, dust hazard	Poverty, marginalization, decreased social and economic resilience, population movements
6	Carbon cycle	Climate change, acidification of ocean surface waters	Shift from fossil fuels to biofuels conflict with food production Shift in growing seasons and risk of crop failure	Respiratory diseases related to air pollution	Risk of flood related damage to property, in coastal and riverine areas	Upto 80 per cent of energy supply is derived through manipulation of the carbon cycle
7	Nutrient cycles	Eutrophication of inland and coastal water Depletion of phosphate resources	-	Health effects from bioaccumulation of N or P in food chains Non-potable water	–	Benefits of food security and biofuel production
8	Acidifying cycles	Acid depositions and drainage land & water ecosystems Acidification of ocean and fresh waters	Freshwater fish resources declining; risk of further collapse of marine fisheries	Poisoning from increased plant and animal uptake of toxic metals	–	Economic damage to forest, fisheries and tourism Corrosion of infrastructure and industrial facilities

Forest are not just trees, but part of ecosystems that underpin life, economies and societies. Where forests are privately owned, they are often managed mainly for production. Yet, in addition to directly supporting such industries as timber, pulp and biotechnology, all forests provide a wide range of ecosystem services. These services include prevention of soil erosion, maintenance of soil fertility, and fixing carbon from the atmosphere as biomass & soil organic carbon. Forest has a large proportion of terrestrial biodiversity, protect water catchments and moderate climate change. Forest also support local livelihoods, provide fuel, traditional medicines and foods to local communities.

Between 1990 and 2005, the global forest area shrank at an annual rate of about 0.2 percent. Losses were greatest in Africa, Latin America and the Caribbean. However forest area expanded in Europe and North America. In Asia and the Pacific, forest area expanded after 2000.

In addition to the changes in global area, significant changes also occurred in forest composition, particularly in the conversion of primary forest to other types of forests (especially in Asia and the Pacific).

APPENDIX 2.2

Nature of pollutants in soil

S. N	Sources of Soil Pollutions	Gases	Colloids	Suspended particles	Dissolved cations	Dissolved anions
1	Soil	CO_2	Clay, Fe_20_3, $Mn0_2$	Clay, sand, silt	N^{a+}, k^+, Ca^{2+}, Mg^{2+}, Mn^{2+}, Co^{2+}, Fe^{3+},	CO_3^{2-}, HCO_3^-, OH^-, Cl^-, SO_4^{2-}, F^-, HSO_3^-
2	Decomposed organic matter	SO_2, H_2, NH_3, CH_4, CO_2	Organic waste materials	Humus organic waste	H^+, Na_+, $NH4^+$	Organic radicals, NO_3^-, NO_2^-, SO_4^{2-}, Cl^-
3	Soil organisms	–	Algae, Fungi, bacteria, protozoa, viruses, ascaris etc.	Algae, bacteria	–	–

Note: Pollution of land is caused by disposal of solid waste and refuse from domestic, Industrial & agricultural sectors.

Industrial wastes are:

- Chemicals residues
- Fly ash from thermal power station
- Plastics
- Rubber
- Glass
- Discarded metal

Agricultural residues are:

- Pesticides
- Fertilizers

Another major sources of pollution not known to the general public is the creation of derelict land due to mining. Roughly 0.8 mha of land in India are despoiled due to open or surface and underground mining activities. Though this problem is highly location-specific and restricted to remote areas, it necessary warrants attention in terms of wise management of land. Somewhere someone is affected due to mining, and reclamation of such derelict land is needed.

For Proper Management of Land

- Wastelands should be afforested on a massive scale involving local people. People themselves should select trees that will meet their requirements.

- The demand for timber should be drastically reduced. Substance furniture material and packing cases should be used. This would ease pressure on standing forests.
- Catchment areas of watersheds must be thickly vegetated. This would hold rain water and recharge springs, rivers, etc.
- Cultivation on hilly slopes should require terracing and building along contour lines.
- Tanks should be desilted, check dams constructed and small ponds created to hold run-off water.
- Shifting agriculture should be replaced by settled agriculture.
- Fertilizers and micronutrients should be applied correctly and only if required. Periodic sampling of soils should be done.
- The drainage problems of field should be attended to.
- Soil fertility should be restored by using cultural practices like mulching.
- Green manuring should be done.
- Introduction of leguminous crops.
- More use of organic manure.

  Excessive application of inorganic fertilizers is not a healthy way of practicing agriculture. Traditional methods of multiple cropping and intercropping to maintain soil fertility have to be given more emphasis. Cereal crops can be mixed with nitrogen fixers and grown together e.g., maize and beans.
- Grazing of cattle in forests must be checked. Rotational grazing and hand cutting of grass will save pasture lands. Stall feeding has to be implanted. Creation of more pasture lands and reclamation of wastelands for pasture development through propagation of new grasses and application of optimum fertilizers need to be implemented.
- Most important, local communities must be reeducated on the need to have the dung alone, rather than burning it for fuel or selling for cash. It well regenerates the land, paying rich dividends in the long run.
- The location of industries must be carefully studied. Industries, like thermal power station and dams, should not displace prime agriculture land.
- The unplanned or haphazard growth of urban development must be checked. 35% of the world's population lives in cities.

  The solutions to our problems are not easy and straight forward but they are complex and inter-connected. As an individual one can contribute in one's own small way. The solutions can be tackled as follows:
- First and foremost, be prepared to act, analyse weather you are contributing to any problems. For example, if you have been throwing garbage on the streets, don't do it. Put in the bin.
- Farmers must be educated on the right agricultural methods to be adopted.
- A village can be adopted to take up a community development project as a project.
- A few saplings may be planted and maintained by the villagers. If these are commercial trees, the villagers will have a greater incentive.
- Collective participation in water and land management programs by:
  - Tree planting on a large scale on barren land.
  - Contour building.

- Construction of small water reservoirs.
- Control of grazing.
- Desilting of tanks.
- Influencing the government on major issue such as the location of heavy industries that would displace or disrupt the lives of the people.

Soil condition indicators are tabulated below.

Table 2.2T_1: Soil condition indicators

S. No.	Indicator	Degradation process
1	Erosion – Sheet	Sheet erosion is caused by rain splash and diffuse water flows. It removes topsoil and reduces productivity, terrestrial biodiversity & ecosystem functions. Many soil have eroded severely in the past to the extent that topsoil has been completely removed. Off-site sediment and nutrient export affects water quality, aquatic ecosystem function & productivity.
2	Erosion-gully	Gully erosion is the erosion of topsoil and subsoil by concentrated overland water flow. It reduces land management options, water quality and terrestrial & aquatic ecosystem function through the delivery of sediment. Gully erosion is expected to be sensitive to climate change.
3	Erosion – wind	Wind erosion reduces air quality, land management options and terrestrial ecosystem function. Burial and deposition of nutrients can harm biodiversity. Dust deposition can significantly alter aquatic ecosystems. Wind erosion is expected to be sensitive to climate change.
4	Acidity	Soil pH is a major indicators of soil chemical health. Acidification can reduce soil health & productivity and ecosystem function. Acidity is associated with erosion, soil structure decline and salinity.
5	Organic carbon	Organic carbon is a prime biological determinant of soil health. Organic carbon comes from the atmosphere. Soil has the largest concentration in the carbon cycle.
6	Soil structure	Soil structure is the arrangement of soil particles and voids. It governs soil water storage and movement & gas exchanged and is the prime physical determinant of soil condition. Soil structural condition is sensitive to land management pratices.
7	Coastal acid sulfate soils	Coastal acid sulfate soils are low-lying coastal soils from previous marine environments which, when drained or exhumed, can discharge sulfuric acid. They have the potential to cause profound terrestrial and aquatic ecosystem damage.

8	Inland acid sulfate soils	Sulfide sediments have often been through of only as a coastal phenomenon but are now known to be common in inland wetlands.
9	Soil salinity	Soil salinity is the accumulation of salt on near the ground surface. It has the potential to cause profound terrestrial and aquatic ecosystem damage, including massive erosion.

Appropriate land management is vital for the sustainable use of soil and land resource. The physical capacity of the land to sustain a range of land uses and management practices in the long term should resist the degradation of soil, and, air & water resources. Failure to manage land sustainably may result in a degradation of resources, and a decline in productivity, natural ecosystem value & infrastructure.

APPENDIX 2.3

Persistence level of various pesticides in soil

S.N	Pesticides	Persistence level*, years
1	Chlordane	5
2	DDT	4
3	Dieldrin	3
4	Picloram	1.5
5	2,4,5-T	0.5
6	2,4-D	0.1

* Time needed for pesticides level to be reduced to <25% of the original level applied.

Note: The use of certain agrochemicals has detrimental effects associated with some environmental ecological damages as outlined below:

(1) Fertilizers:

(i) Groundwater contamination: Excessive use of fertilizers can lead to the contamination of groundwater with nitrate, rendering it unfit for consumption by humans or livestock. Water containing large concentrations of nitrate can poison animals by immobilizing some of the hemoglobin in blood, reducing the ability to transport oxygen.

(ii) Eutrophication: In addition, the run-off of agricultural fertilizers into streams, lakes and other surface water can cause an increased productivity of those aquatic ecosystems, a problem known as eutrophication. It starts by increasing the amount of phytoplankton in the water, this then blocks out the sunlight, which leads to plants being unable to photosynthesize, which in turn means that the fish & other creatures don't have anything to eat, which means they die and then in turn so do other animals that depended on them such as birds.

(2) Pesticides: The use of pesticides raises a number of environmental concerns. Over 98% of sprayed insecticides and 95% of herbicides reach a destination other than their target species, including non-target species, air, water and soil.

(i) Air pollution by pesticides

Pesticides can contribute to air pollution. Pesticides drift occurs when pesticides suspended in the air as particles are carried by wind to other areas, potentially contaminating them. Pesticides that are applied to crops can volatilize and may be blown by winds into nearby areas, potentially poising threat to wildlife.

(ii) Water pollution by pesticides

There are four major routes through which pesticides reach the water:

(a) It may drift outside of the intended area when it is sprayed,

(b) It may percolate, or leach, through the soil,

(c) It may be carried to the water as runoff or it may be spilled, for example accidentally or through neglect.

(d) It may also be carried to water by eroding soil.

There are some factors that also affect pesticide's ability to contaminate water and these include:

- Its water solubility
- The distance from an application site to a body of water
- Weather
- Soil type
- Presence of a growing crop
- The method used to apply the chemical.

(iii) Soil pollution by pesticides: Many of the chemicals used in pesticides are persistent soil contaminants, whose impact may endure for decades and adversely affect soil conservation. The use of pesticides decreases the general biodiversity in the soil. Not using the chemicals result in higher soil quality with the additional effect that more organic matter in the soil allows for higher water retention. This helps to increase yields for farms in drought years. When organic matter in the soil increases the amount of pesticide that will leave the area of application because organic matter binds to and helps break down pesticides.

Effect of pesticides on plants:

- It hinders nitrogen fixation in legumes. The insecticides DDT, methyl parathion, and especially pentachlorophenol has been shown to interface with legume rhizobium chemicals signaling. Reduction of this symbiotic chemicals signaling result in reduced nitrogen fixation and reduced crop yields.
- Pesticides can kill bees and they are strongly implicated in pollinator decline and the loss of species that pollinate plants.
- On the other side, pesticides have some direct harmful effects on plant including poor root hair development, shoot yellowing and reduced plant growth.

Effect of pesticides on animals:

Pesticides inflict extremely widespread damage to biota, and many countries have acted to discourage pesticide usage through their Biodiversity Action Plans.

(i) Animals may be poisoned by pesticide residues that remain on food after spraying, for example when wild animals enter sprayed fields or nearby areas shortly after spraying.

(ii) Widespread application of pesticides can have eliminated food sources that certain types of animals need, causing the animals to relocate, change their diet, or starve.

(iii) Poisoning from pesticides can travel up the food chain; for example, birds can be harmed when they eat insects and worms that have consumed pesticides.

(iv) Some pesticides can bio accumulate, or builds up to toxic levels in the bodies of organisms that consume them over time, a phenomenon that impacts species high on the food chain especially bird.

(v) Some pesticides come in granular form, and birds & wildlife may eat the granules, mistake them for grains of food. A few granules of a pesticides are enough to kill a small bird.

(vi) The herbicide parquet, when sprayed onto bird eggs, causes growth abnormalities in embryos and reduces the number of chicks that hatch.

Uses and purpose of various classes of Pesticides are tabulated below.

Table 2.3T_1: Uses and purpose of various classes of Pesticides

S.No.	Classes of pesticides	Uses and purpose
1	Acaricide	Used against mites and ticks, members of the acaridae.
2	Algaecide	Used against algae.
3	Antifouling agent	A paint additive to protect against organisms that grow on moist or wet surfaces including those underwater.
4	Attractant	A chemical that attracts insects or birds or other animal pests to a trap or a poison
5	Avicide	Kills or discourages birds.
6	Chemosterilant	A chemical that will lower or stop reproduction by a pest.
7	Defoliant	A chemical that induces plants to drop their leaves prior to harvest (sometimes used as a herbicides or silvicide)
8	Dessicant	A chemical that dries plants out before harvest.
9	Fumigant	Used to kill soil pests and sometimes weeds.
10	Fungicides	Used against fungi.
11	Herbicides	A weed killer; defoliants & silvicides are often called herbicides.
12	Insecticides	Used against insects.
13	Miticides	Used against mites and ticks
14	Molluscide	Used to kill or control mollusks and other invertebrates (eg, predatory & nails slugs, oyster drill, lamprey)
15	Nematicide	Used against nematodes; tiny round worms that feed on decaying matter, roots or other parts of plants.
16	Ovicide	A chemical used against the eggs of insects, mites & nematodes.
17	Piscicide	Used to reduce the population of rough fish in a body of water.
18	Plant regulator	A chemical that alters the normal pattern of growth & development of a plant; usually used for herbicidal purpose.
19	Repellant	A chemical that repels insects or other rodents.
20	Rodenticides	Used against mice, rats and other rodents.
21	Silvicide	Used against bush, trees, shrubs & other woody plants (often lumped with herbicides)

As far as human beings are concerned, the harmful effects of insecticides have been a cause of concern. The people working on farms, involved in the handling, etc., are exposed directly and are likely to become victims of the toxic effects of insecticides which may be lethal. Organophosphates decompose more rapidly than the organochlorines but some of them have a toxic strength which is 300 times higher. Organophosphates build up residues on foliage after application. The toxicity ratings of some of the insecticides are given in Table below.

Table 2.3T$_2$: Toxicity ratings of some common organic phosphorous insecticides

S. N	*Common or trade name*	Oral LD$_{50}$ (rates, mg/kg)
1	Malathion	1500
2	Methyl parathion	25
3	Parathion	10
4	Chlorthion	1500
5	Nilate or Ethion	200
6	Guthion	1500
7	Phostex	1200
8	Trithion	100
9	Dibrom	430
10	Dementon or Systox	10
11	Di-Syston	10
12	Phosdrin	12

Note: Data on oral toxicity or rates. LD$_{50}$ values represent the amount of insecticide per unit of body weight required to cause the death of 50% of group of test animals.

Persistence of insecticides in the environment after application in not desirable from environmental consideration. In general, the organochlorines persist longer in the environment than the organophosphorus based insecticides. Data regarding persistence of organochlorines in the soil are given in Table below.

Table 2.3T$_3$: Approximate persistence of organochlorines insecticides in soil

S.N	Insecticides	Duration of persistence, years
1	Toxaphene	> 6
2	Heptachlor	> 9
3	Aldrin / Dieldrin	> 10
4	DDT	> 10
5	Hexachlorocyclohexane	> 10
6	Chlordane	> 12

Relative insecticide toxicities of some organochlorine insecticides is tabulated below.

Table 2.3T$_4$: Relative insecticide toxicities of some organochlorine insecticides

S.N	Insecticides	Class of insecticides	Relative toxicity, Oral LD_{50} values, mg/kg or ppm
1.	Aldrin	Organochlorine	39
2.	Dieldrin	Organochlorine	46
3.	DDT	Organochlorine	113
4.	Carbarly	Carbonate	540
5.	Methoxychlor	Organochlorine	6000

DDT affect vegetables, fish, milk, etc. There have been reports that even breast fed rabies get 70% more than the accepted amount of DDT. Like human beings, other species are also affected adversely in various ways, e.g., thinning of egg shells, etc., Therefore, precautions have to be taken for the presence of pesticide residues on foods to ensure that the quantities remaining on a fruit, vegetable, grain or animal product are reduced to such level that do not pose any health hazard to human beings.

There is a definite trend towards finding suitable alternative for those insecticides which have adverse effects on the environment. In general, the organophosphorus based insecticides show greater promise than organochlorine based ones. The harm caused by DDT has been well-established hence most countries including India has banned the use of DDT.

Table below shows the tolerance limits for pesticides residues.

Table 2.3T$_5$: Tolerance limit for pesticides residue

S.N	Trade name	Chemical name	Tolerance limit in ppm	Loads
1	DDT	1, 1, 1,-trichloro-2, 2-bis (p-Chloro phenyl)- ethane	7	Apples, beets, cattle fat, peas, etc.
2	Dieldrin	1, 2, 3, 4, 10-hexacholoro-6, 7-epoxy-1, 4a, 5, 6, 7, 8a-octahydro-1 endoexo-5, 8-dimethanonapthalene	0.1	Carrot, Grapes, oats, potatoes, etc.
			0.25	Apples, beans, celery, peas, strawberries, etc.
3	Malathion	0, 0-dimethyl 1, 5-(1, 2-dicarbomethoxyethyl) phosporodiathioate.	4.0	Meat of cattle, hogs, poultry and sheep.
			8.0	Apples, beans, celery, peas, strawberries, etc.
4	Pyrethrums	Pyrethrums	1.0	Post-harvest application on apples, beans, peas etc.
			3.0	Corn, rice, wheat, etc.

Handling of Pesticides

The harmful effects of pesticides depend on the following factors:

- Physical form of the pesticides used.
- Purity of the active material in the pesticides used.
- Absorption rate of the pesticide in the body of human beings.
- Detoxification and distribution of pesticides in the body.
- Solubility of the pesticide in the body fluid

Analysis of typical soil samples irrigated by different water/effluent is tabulated below.

Table 2.3T_6: Analysis of typical soil samples irrigated by different water/effluent

S.N	Soil irrigated by	pH value	Conductivity	Organic matter	Nitrate nitrogen	Phosphorous	Ca, Mg
1	Natural rain water	7.5	0.3	0.3	6.3	9.6	24
2	Well water	8.1	0.3	0.9	6.7	12.0	20.8
3	Waste water from paper mills	8.2	0.5	0.7	7.8	12.0	30
4	Effluents from pulp factory	8.1	0.6	1.2	7.3	20.0	28.8
5	Effluent from batch pulp & paper plant	8.5	0.5	1.2	8.1	9.6	22.4

Classification of pesticides with example are tabulated below:

Table 2.3T_7: Classification of pesticides with example

S.N	Classification of pesticides	Example of pesticides of various classes
1	Chlorinated hydrocarbons	Aldrin, Chlordane, BHC, DDT, Dieldrin, Lindane, Heptachlor (PCB)
2	Organo phosphorous compounds	Malathion, Parathion (methyl & ethyl), Guthion
3	Carbonate compounds	Zectrion, Carbonyl (sevin)
4	Inorganic compounds	$CuSO_4$, $NiCl_2$, PbO_2, As_20_3
5	Miscellaneous compounds	2,4,D,2,4,5,T carnayl organic mercurials

Level of chlorinated hydrogen pesticides in body fat in different countries (ppm) is given below.

Table 2.3T_8: Level of chlorinated hydrogen pesticides in body fat in different countries (ppm)

S.N	Country	DDT	DDE	BHC	Dieldrin	Endrine	Heptachlor
1	U.S.A	3.7	6.9	0.40	0.25	0.02	0.10
2	France	3.1	6.5	1.19	-	-	-
3	India	18	12	1.43	0.04	0	0
4	U.K	-	-	0.02	0.21	0.02	0.01

TLV, LD_{50} Dose & % degradation (at various temperature) of different pesticides is tabulated below:

Table 2.3T_9: TLV, LD_{50} Dose & % degradation (at various temperature) of different pesticides

S.N	Pesticides	Threshold limiting values (TLV), mg/m³	LD_{50} dose, mg/kg	% degradation at various temperatures		
				600°C	800°C	1000°C
	Insecticides:					
1	Aldrin	0.25	54-56	-	-	-
2	PCBs	0.50	250	-	-	-
3	Carbonyl	-	340	-	-	-
4	DDT	1	420-800	99.2	99.7	100
5	Eldrin	-	5-43	-	-	-
6	Heptachlor	0.5	90	-	-	-
7	Lindane	0.5	90	-	-	-
8	Malathion	10	480-1500	95.3	96.3	96.7
9	Parathion	0.1	4-30	64	96.5	97.5
10	Dieldrin	0.25	40	99	99.5	99.5
	Herbicides:					
11	2, 4 D	10	666	-	-	-
12	2, 4, 5 T	-	0.03	-	-	-
13	Pentachlo-rophenol	0.5	27-80	-	-	-

14	Dioxane	-	0.03	-	-	-
	Fungicides:					
15	Copper sulfate	-	Very high	-	-	-
16	Pharsenol	-	825	-	-	-
17	Picloram	-	-	90.8	95.6	99.2
18	Atrazine	-	-	87.8	88.8	89.0
19	Sevin	-	-	88.7	88.8	90.0
20	Dapalon	-	-	64.3	64.7	97.0
21	DN BP	-	-	99.8	99.8	99.8
22	Nemagon	-	-	99.6	99.6	99.6

Fresh water permissible limits of various pesticides is tabulated below.

Table 2.3T$_{10}$: Fresh water permissible limits of various pesticides (mg/litre)

S.N	Class of pesticides	Trade name/ Chemical name of pesticide	Structural formula	Uses	Permissible limit fresh water, mg/ litre
1.	Chlorinated hydro-carbons	(a) Aldrin-Dieldrin	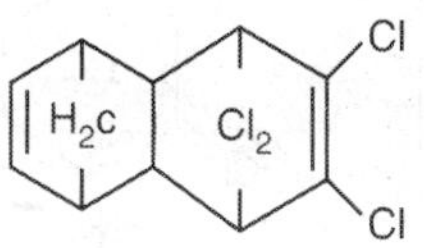	Soil insecticide for control of ants, beetles & cotton pests. Its use, is now banned in U.S.A	0.003
			Aldrin 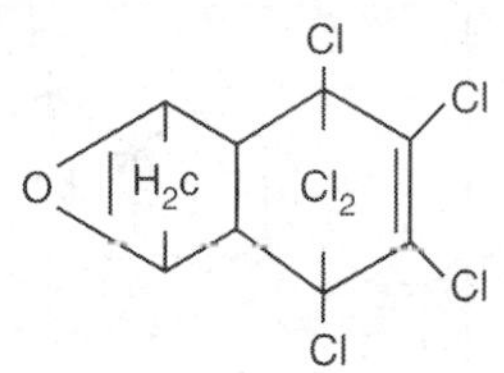	-do-	-do-
		(b) chlordane	Dieldrin 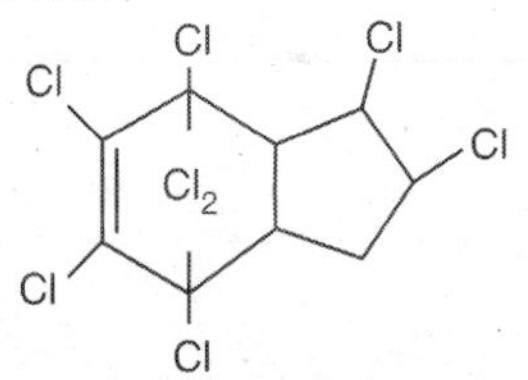	Effective against termites, potential carcinogen, use banned in U.S.A since 1975	0.01

		(c) Lindane (1,2,3,4,5,6 hexachloro -cyclohexane)	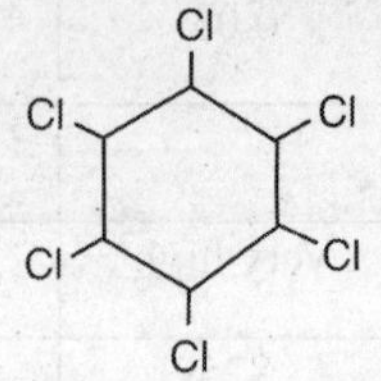	Control of cotton insects and rice stem borer	0.01
		(d) DDT [1,1,1-trichloro-2, cl 2-bis (p-Chlorophenyl ethane)]	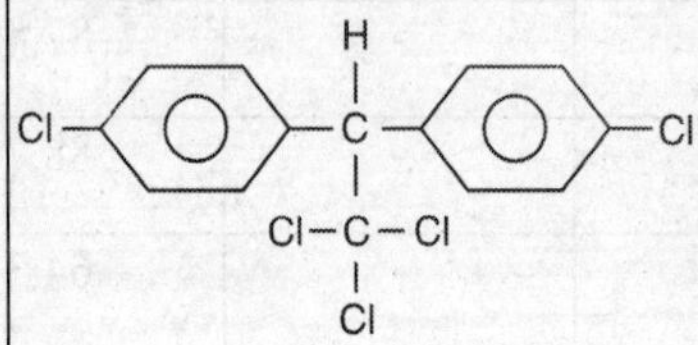	Broad spectrum- Cotton, soyabean, and peanut pastes, mosquito control, persistent in environment, accumulated in food chain, use banned in U.S.A	0.001
		(e) Toxaphene	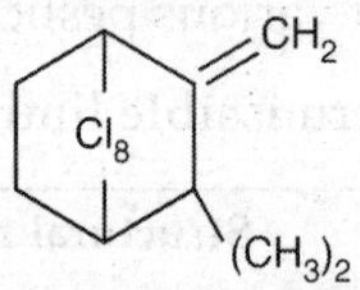	Insect control on crops and livestock, widely used in U.S.A, ban has been proposed for carcinogenic properties	5
		(f) Heptachlor	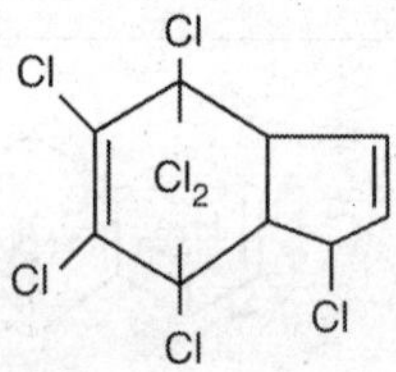	Pests control in soil, use suspended due to potential carcinogenicity	0.001
		(g) Endrin	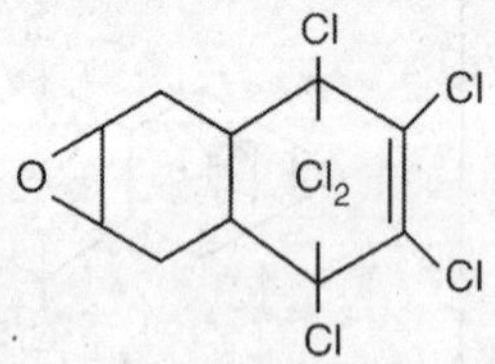	Effective against black current mud-mite, also used as a zoocide, precaution to be taken to avoid skin contact during application	0.004

		(h) Methoxychlor	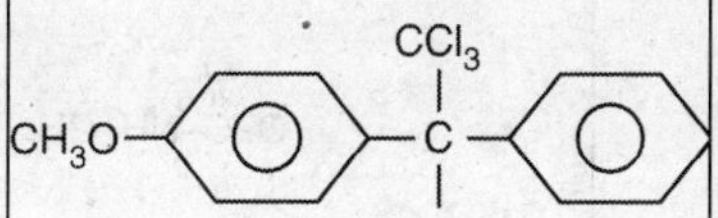	It is a popular DDT substitute, reasonably biodegradable, low toxicity to mammals	0.03
2	Organophosphates	(a) Malathion [o,o dimethyl-5 (1,2 dicarboxyethyl) phosphorodithioate]	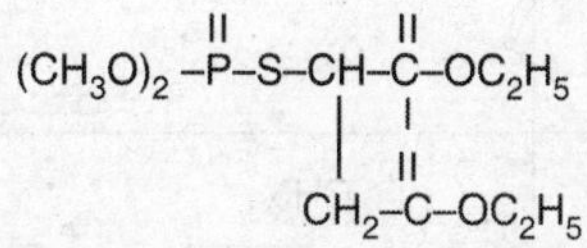	Controls some pests of fruit and vegetable- little hazards to mammals	0.1
		(b) Parathion [0,o diethyl o-p nitro phenyl phosphorothionate]	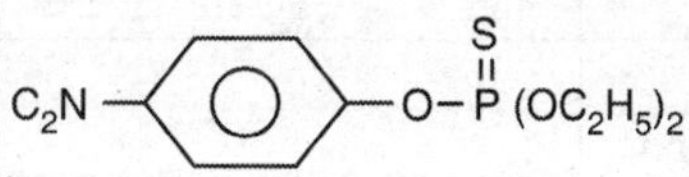	Larvicide for mosquito control, also broad spectrum insecticide for fruit and vegetable pests.	0.1
		(c) Methyl parathion	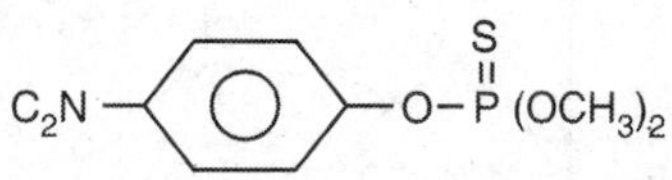	Control of plant pests, ranks second in pesticide consumption in U.S.A	—
		(d) Diazinor	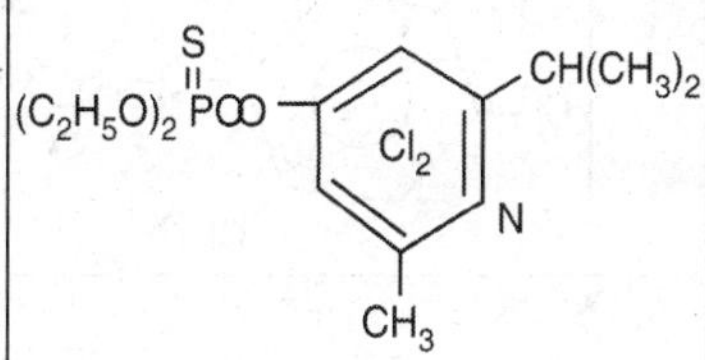	Control of many vegetable and fruit pests	—
3	Carbonates:	(a) Baygon	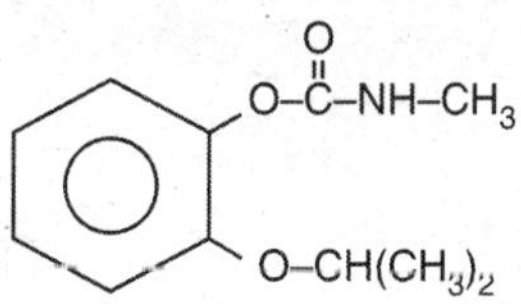	Control of flies, mosquitoes, ants & cockroaches	—

		(b) Carbonyl (sevin) [1, naphthyl- n-methyl carbonate]	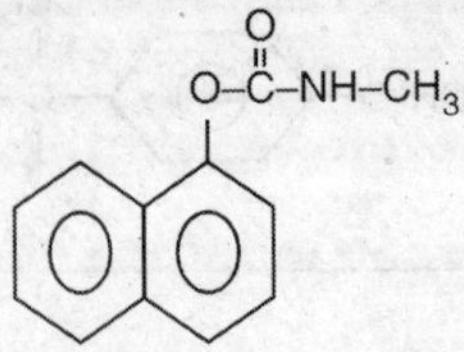	Used on cotton crops, forage, fruits and vegetables; Lawn and garden insecticides; Low toxicity to mammals	—
		(c) Dimetilan	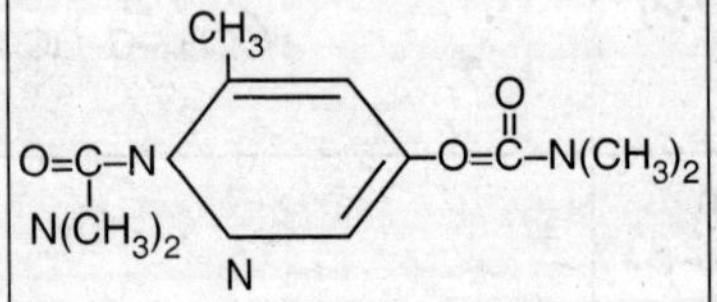	Control of house and fruit flies	—
4	Chloroph- enoxyacids:	(a) 2,4-D [2,4 diclorophenoxy acetic acids]	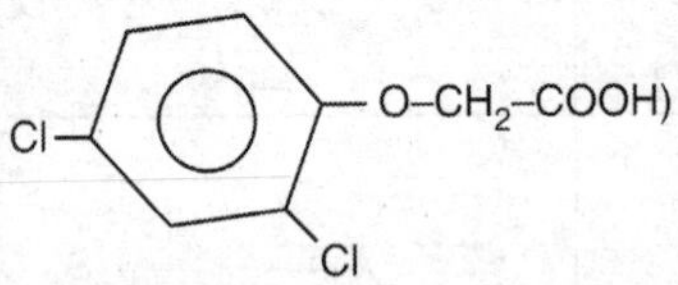	Herbicide control of brand leaved weeds, aquatic vegetation, military defoliant (may contain highly toxic TCCD as impurity)	100
		(b) 2,4,5-T [2,4,5 dichloropheoxy acetic acid]	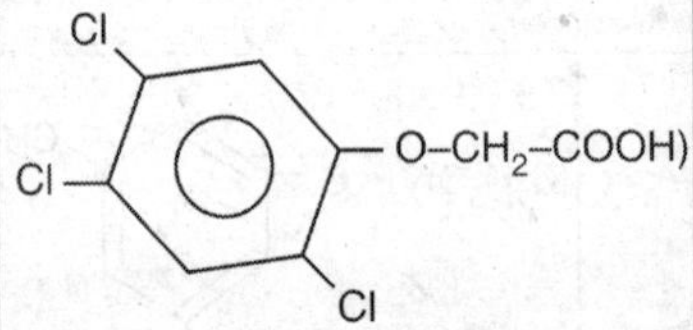	Weed control, military defoliant	

APPENDIX 2.4

Status of soil pollution in India

Salient features of soil contamination in India is as follows:

- There are >125 major contaminated sites across the country.
- 175 million hectares of land (out of 329 million ha) are considered degraded.
- > 40% of chemical fertilizers leached into soil.
- 14 States are affected by Fluoride contamination.
- > 65 per cent of India's villages are exposed to residual pesticides risk.
- Heavy metals beyond permissible limits affecting ground water of 40 districts from 13 states.

Identification of the Priority Sites of land contamination:

The priority sites have been identified based on following criteria:

- Constituents of contamination.
- Ground water contamination.
- Affected population/ flora/ fauna.
- Health and Ecological impacts.

To undertake remediation, 23 sites across 12 States identified as priority sites.

Priority list of HW (Hazardous waste) contaminated Dumps Sites in India is tabulated below.

Table 2.4T$_1$: Priority list of Hazardous waste contaminated Dumps Sites in India

S.N	State	Name of the Area	No. of Site	Public/private Sector	Nature of Contaminant
1	Gujarat	Vadodara	1	Private Sector Unit	Chromium
2	Kerala	Eloor-Edayar, Cochin	4	Public/private Sector Units	Heavy metal and POPs
3	Madhya Pradesh	Ratlam	4	private Sector Units	Gypsum, iron salts and Naphthalene
4	Orissa	Ganjam	4	Private Sector Unit	Mercury
5	Orissa	Talcher	1	Public Sector Units	Chromium
6	Orissa	Sundergarh	4	Private Sector Units	Chromium
7	Rajasthan	Bichhadi	1	Public/Private Sector Unit	Inorganic Salts, Organics
8	Tamil Nadu	Ranipet	1	Public Sector Unit	Chromium
9	Uttar Pradesh	Rakhimandi, Kanpur	1	Unknown	Chromium

10	Uttar Pradesh	Rania, Kanpur Dehat	1	Unknown	Chromium
11	Uttar Pradesh	Lucknow	1	Private Sector Unit	HCH (hexa Chloro cyclo hexane)
12	West Bengal	Nibra Village, Howrah	1	Unknown	Chromium

POPs- Persistent Organic Pollutants

Process Involved in contaminated land Sites Management:

- Notification of the contaminated site.
- Initial survey.
- Reporting of the initial/preliminary investigations
- Detailed site characterization.
- Risk assessment.
- Submission of a detailed site investigation Report.
- Finalization of Remedial measures for soil and groundwater.
- Preparation of DPR (Detail Project Report).
- Public Hearing.
- Implementation of remediation technology.
- Site closure options/ future land use patterns.

Possible Remediation Methodologies

- Transfer of contaminated solid wastes to common TSDFs (Transfer station disposal facilities) or possible options for utilization of the waste removed.
- Capping the waste in a secured landfill (SLF).
- Confinement of the contaminated area by concrete/ bentonite side walls and capping.
- Contaminated soil walls and capping.
- Contaminated soil excavation, soil washing and refilling.
- Physical and chemical remediation techniques like Soil vapour extraction, pump & treat, chemical precipitation etc.
- Bioremediation of contaminated sites.

Possible Remediation Outcomes include improved:

- Quality of ground/surface water and soil.
- Productivity and better land use pattern.
- Human health, flora/fauna.
- Learning to support future projects.
- Awareness of industries/ local communities.
- Capacity Building of environmental regulatory agencies.
- Recuperation of land for future use.

Provision of the National Clean Energy Fund (NCEF) in rupees for remediation

- Assessment of contamination, design of remediation plan, bid document preparation and supervision of remedial activities:
  - 60 Crore (5 crore per site for 12 sites)

(iii) Remediation & Monitoring Of Hazardous Wastes dumpsites:
  - 720 Crore (crore per site for 12 sites).

- Regular monitoring of remaining contaminated sites:
- 25 Crore (49 lakh per site)
- Government of India (GOI) to provide 40% of cost.
- State Governments to provide 30% of cost.
- 30% funds from the identified polluter under the Polluter Pays Principal or Public Private Partnership.

Biological Treatment of contaminated soil

- Phytoremediation especially suited for heavy metals.
- Bioremediation for organic pollutants.
- Land farming as extended application of Bioremediation.

Conventional Methods Vs Bioremediation for soil decontamination:

Conventional Methods

- Costly.
- Create new wastes.
- Does not eliminate problems.
- Low public acceptance.

Bioremediation technology

- Cost-effective.
- Eliminated problem to a greater extant.
- Generate no or low waste.
- High Public acceptance.

Note: Causes of Land Pollution

Below are the causes of land pollution:

- **Deforestation and soil erosion:** Deforestation carried out to create dry lands is one of the major concerns. Land that is once converted into a dry barren land, can never be made fertile again, whatever the magnitude of measure to redeem it are. Land conversion, meaning the alteration or modification of the original properties of the land to make it use worthy for a specific purpose is another major cause. This hampers the land immensely. Also there is a constant waste of land. Unused available land over the years turns barren; this land then cannot be used. So in search of more land, potent land is hunted and its indigenous state is compromise with.
- **Agricultural activities:** With growing human population, demand for food has increased considerably. Farmers often use highly toxic fertilizers and pesticides to get

rid of insects, fungi & bacteria from their crops. However, with the overuse of these chemicals, they result in contamination and poisoning of soil.

- **Mining activities:** During extraction and mining activities, several land spaces are created beneath the surface. We constantly hear about land caving in; this is nothing but nature's way of filling the spaces left out after mining or extraction activity.
- **Overcrowded landfills:** Each household produces tonnes of garbage each year. Garbage like aluminium, plastic, paper, cloth, wood is collected and sent to the local recycling unit. Items that cannot be recycled become a part of the landfill that hampers the beauty of the city and cause land pollution.
- **Industrialization:** Due to increase in demand for food, shelter and house, more goods are produced. This result in creation of more waste that needs to be disposed off. To meet the demand of growing population, more industries were developed which led to deforestation. Research and development paved the way to modern fertilizers & chemicals that were highly toxic and led to soil contamination.
- **Construction activities:** Due to urbanization, large amount of construction activities are taking place which has resulted in large waste articles like wood, metal, bricks, plastic that can be seen by naked eyes outside any building or office which is under construction.
- **Nuclear waste:** Nuclear plants can produce huge amount of energy through nuclear fission and fusion. The left over radioactive material contains harmful and toxic chemicals that can affect human health. They are dumped beneath the earth to avoid any casualty.
- **Sewage treatment:** Large amount of solid wastes is leftover once the sewage has been treated. The leftover material is sent to landfill site which end up in polluting the environment.
- **Atmospheric deposition:** Air pollution doesn't remain air pollution forever. Ideally it disperses, so the concentration of problematic chemicals becomes so low that in no longer contributes to pollution. Sometimes, though, it falls back to the ground and becomes either water pollution or land pollution.

Effects of Land Pollution

- **Soil pollution:** Soil pollution is another form of land pollution, where the upper layer of the soil is damaged. This is caused by the overuse of chemicals fertilizers, soil erosion caused by running water and other pest control measures.
- **Change in climate pattern:** The effects of land pollution are very hazardous and can lead to the loss of ecosystems. When land is polluted, it directly or indirectly affects the climate patterns.
- **Environmental impact:** When deforestation is committed, the tree covers is compromised on. This leads to a steep imbalance in the rain cycle. A disturbed rain cycle affects a lot of factors.
- **Effects on human health:** The land when contaminated with toxic chemicals & pesticides lead to problem of skin cancer and human respiratory system. The toxic chemicals can reach our body through food and vegetables that we eat as they are grown in polluted soil.

- **Cause Air pollution:** Landfill across the city keep on growing due to increase in waste and later burned which leads to air pollution. They become home for rodents, mice etc. which in turn transmit diseases.
- **Distraction for Tourist:** The city looses its attraction as tourist destinations as landfill do not look good when you move around the city. It leads to loss of revenue for the state government.
- **Effects on wildlife:** The animal kingdom faces a serious threat with regards to loss of habitat and natural environment. The constant human activity on land, is leaving it polluted; forcing these species to move further away and adapt to new regions or die trying to adjust
- **Extinction Of Species:** Species are pushed towards endangerment and extinction primarily by two processes-Habitat fragmentation and Habitat destruction. Both the actions can cause some species to go extinct and others to become invasive.
- **Bio Magnification:** Process in which certain non-biodegradable substance go on accumulating in the food-chain. It puts all the species above and below it at risk, and ultimately affects the food pyramid.
- **Effects On Biodversity:** Species extinction and magnification is going to overthrow the balance of nature very significantly. The main reason for this is disturbance created in the food chain.

Other issues that we face include increased temperature, unseasonal weather activity, acid rains etc. The discharge of chemicals on land, makes it dangerous for the ecosystem too.

Solutions for Land Pollution

- Make people aware about the concept of Reduce, Recycle and Reuse.
- Reduce the use of pesticides and fertilizers in agricultural activities.
- Avoid buying packaged items as they will lead to garbage and end up in landfill site.
- Ensure that you do not litter on the ground and do proper disposal of garbage.
- Buy biodegradable products.
- Do Organic gardening and eat organic food that will be grown without the use of pesticides.
- Create dumping ground away from residential areas.
  - i. Encourage organic farming.
  - ii. Proper garbage disposal.
  - iii. Recycle garbage.
  - iv. Reduce use of herbicides and pesticides.
  - v. Avoid over packaged items.
  - vi. Efficient utilization of resources and reducing wastage.

Several creatures survive under the land too. Disrupting the harmony of the land, is disrupting their habitat. This has led to several creatures reaching the endangered status like the Gilbert's Potaro in Australia.

We walk and survive on land. It is literally the base of our ecosystem. It is in our good interest to take care of it.

Some important facts on land pollution:

Facts on paper recycling

- Recycling a tonne of paper spares 17 trees, and Recycling half the worlds paper would free 20 million acres of forest land.
- Recycling paper uses 70% less energy than to manufacture paper from virgin materials.
- Every 10 tonnes of paper recycled is 10 tonnes less on its way to landfill.
- Paper and cardboard can decompose fairly easily, but the process gives off methane (a greenhouse gas), which contributes to climate change.

Facts on metal recycling

- Recycling an aluminium can save enough energy to run a TV for 3 hours.
- Once an aluminium can is recycled, it is part of a new can within 6 weeks.
- There is no limit to the amount of times an aluminium can be recycled.
- The energy it takes to make one new aluminum can is the same as it is to make 20 recycled cans.
- Recycling steel and tin cans saves 74% of the energy used to produce them.

General facts

- Land pollutions causes us to lose 24 billion tonnes of top soil every year.
- We throw away enough cans every day to fill 63,000 garbage trucks.
- Over 80% of items in landfills can be recycled, but they're not.
- The main human contributor to pollution are landfills.

APPENDIX 2.5

Recommended Standard Values for Daily intake of Air, Water and Soil by Human Beings

Age (years)	Air Inhalation (m^3/day)	Water Ingestion (Litres/day)	Soil Ingestion (mg/day)	Total soil Adhered, mg/day
0-<0.5	2	F:0/0	35	2200
		NBF:0.2/0.8		
0.5-<5	5	0.2/0.8	50	3500
5-<12	12	0.3/0.9	35	5800
12-<20	21	0.5/1.3	20	9100
>20	23	0.4/1.5	20	8700

Note: Soil/Land Pollution Fact Sheet:

- A majority of the human population lives on land or the lithosphere. If a pollutant gets introduced in the lithosphere, it directly affects the surrounding environment, causing soil or land pollution.
- It impacts the entire food chain and thus turns out be the most hazardous of all types of pollution.
- Untreated soil wastes like sewage, insecticides, pesticides, herbicides, litter contamination, slash and burn-techniques, radioactive wastes, and mining activities lead to soil pollution.
- When inorganic and organic sewage is left untreated, it degrades the soil quality. Synthetic material like plastic, nylon, thermocol, and other complex polymer compounds do not decompose for millions of years. They keep accumulating and there is no possible solution to this problem, expect for treatment & proper disposal.
- Organic wastes, if left untreated, can from leached and other harmful chemicals compounds, cause extensive damage to the soil and its fertility. In such cases, the soil cannot be used for cultivation, unless the sewage is properly treated and disposed off in landfill areas.
- Improper use of pesticides, insecticide, herbicides, and fertilizers, can cause permanent long-term damage to the agricultural land. Soil salinity drastically increases due to the use of chemical compounds on a large scale, and may cause it to become infertile.
- Nuclear detonations release a large amount of nuclear wastes that pollute the land water bodies. The Chernobyl disaster, Hiroshima & Nagasaki nuclear explosion, and the recent-Fukushima disaster, are well-known examples of such incidents.
- Improperly conducted mining activities can destroy the landscape of the place where they are undertaken. A barren wasteland results mostly due to mining activities, and is also considered as a form of land pollution.

Apart from the above-mentioned sources, electronic wastes, acid rain, deforestation, rampant urbanization, and open defecation lead to land pollution.

APPENDIX 2.6

Management of soil pollution

The importance of soil as a natural body derives in large part from its role as an interface between the worlds of rocks, air, water and biosphere.

The concept of the soil as interface means different things at different scales. **At The scale of kilometers,** soil channel waters from rain to rivers and transfer mineral elements from rock to the oceans. They also run over and supply vast amount of atmosphere gases substantially influencing the global balance of methane & CO_2. **At a scale of few millimeters,** soil provides diverse microhabitats for air-breathing and aquatic organism, channels water & nutrients to plant roots and provides surface & solution vessels for thousands of biochemical reactions. Finally, **at a scale of a micrometers and less,** soil provides ordered & complex surfaces, both minerals & organic that acts as templates for chemical reaction and interacts with water & solutes. Also, at micro scale, its tiniest mineral particles from micro-zones of electromagnetic charge that attract everything from bacterial cell wall to proteins conglomerates of water molecules.

Potential environmental pollutants inclusive of soil pollutants have been tabulated below.

Table 2.6T_1: Potential environmental pollutants inclusive of soil pollutants and their examples

S.N	Pollutants	Examples
1	Nutrients	N and P in commercial fertilizers, manures, bio solids, Waste water treatment effluent
2	Agrochemicals	Insecticides, herbicides, fungicides etc.
3	Hazardous organic chemicals	Strong acids and bases
4	Acidification	Acid precipitation, acid mine drainage
5	Salinity or Sodicity	Saline irrigation water, salt water intrusion
6	Trace elements	Heavy metals, elements normally present at low concentration in soil and plant
7	Sediments	Eroded soil in surface waters
8	Particulates	Dust (from wind erosion), volcanic dust ash, fly ash
9	Greenhouse gases	CO_2, CH_4 and other relatively active gases
10	Smog form of compounds	Ozone, secondary products of fuel combustion

Mechanism of pollution Transport through soil

The environment plays a key role in the ultimate fate and transport of contaminants. The specific fate of contaminant following their release into the environment, depends on their chemical structure which is highly variable, abiotic factors within the receiving environment (e.g. organic carbon, pH, surface water) and interaction with the biotic environment which can result in degradation, transformation or bio concentration of the contaminants. Once contaminants also reach the soil, they move in one of these directions:

- Vapourized into atmosphere without chemical change.
- Absorbed by soils move downward through the soil in liquid or solution from and can be lost from the soil by leaching.
- Undergo chemical reactions within or on the surface of the soil.
- Broken down by soil microorganisms.
- Wasted into streams and rivers in surface run-off.
- Taken up by plants or soil animals and move up the food chain.

Transport Mechanisms

Channels or contaminants are transported through soil principally by three mechanisms:

- Mass flow of dissolved chemicals within moving solution.
- Liquid diffusion within soil solution
- Gaseous diffusion within soil air voids.

Mass flow: Refers to the passive transport of dissolved solutes within solution by intermolecular collision which moves the solutes from regions of higher solute density to lower solute (concentration).

Vapor diffusion: Chemicals vapor molecules in the soil air spaces also undergo molecular collision and spread out by vapour diffusion which is expressed as the product of the vapour density or concentration gradient and a proportionality coefficient called the soil vapour diffusion coefficient.

Loss Pathways Of Contaminants In Soil

Contaminants added to a soil from the surface may leave the zone of incorporation by one of these three loss pathways:

- **Leaching:** It takes place principally by mass flow and refers to the downward movement of the contaminants.
- **Volatilization:** Refers to the loss of contaminants in vapour form to the atmosphere through the soil surface.
- **Degradation:** Refers to the biological or chemical transformation of the contaminant to a different form with properties distinct from those of the contaminant prior to transformation.

(Subsurface) Transformation Processes

Transformation processes change in a substance of a compound which might result in one more of the following:

- **Detoxification**: An irreversible change in a substance from a toxic to non-toxic from. This occurs most commonly when an organic substance break down into its inorganic constituents with water and CO_2 being main by-products.
- **Transtoxification:** Inverse the conversion of one toxic compound to another toxic compound. In the process, toxicity may remain the same, increase or reduce.
- **Toxification:** Is the conversion of non-toxic compound to a toxic substance.

TOXIC ELEMENTS IN THE ENVIRONMENT

Included among the toxic elements are heavy metals such as Ag, Cd, Cr, Co, Cu, Fe, Hg, Mo, Ni, Pb, Sn, Zn as well as lighter elements such as Al, Ag and Se. Under certain environmental conditions, these elements may accumulate to a toxic concentration and cause ecological drainage.

Natural sources of toxic elements:

- Naturally occurring contaminations, natural emission of toxic elements to the atmosphere can take place by volcanic outputs or outgassing, and by vapour phase of relative volatile elements such as As, Hg and Se. We can have contamination from metalliferous sites.
- **Anthropogenic sources of toxic elements:**

  Agricultural practice: From long-term use of pesticides, sewage sludge (ie by-product of secondary treatments of municipal sewage).

  Pollution from (metal) mining and processing: Pollution from toxic elements in association with various aspects of mining and processing industries. Mining produces ore as a product along with a considerable amount of metal contaminated wastes. Pollution around mines is caused by dumping of contaminated overburden, excavating wastes etc.

  Oil pollution: The contaminants of soil and ground water with mineral oils, hydrocarbons or mineral oil-based produce are among the most common negative effects of the industrial society.

MAJOR CONSEQUENCE OF CHEMICALLY DEGRADED SOILS

- Environmental pollution / contamination.
- Bioaccumulation in plants and animals tissue.
- Accumulation of heavy metals.
  (a) Narrow Concentration Range- The margin between beneficial and harmful is narrow.
  (b) Food Chain – Plants, animals and humans.

Prevention of Inorganic Chemical Contamination

- Reduces the applications to soil.
- Immobilization of the toxins – maintain pH> 7.0.
- Removal of toxins – Remediation ("Clean up").

Physical Remediation by:

- Incineration
- Vacuum Extraction
- Soil washing / Flushing
- Leaching
- Heating

Biological Remediation (Bioremediation) by:

- **Microbial Action**
- **Phytoremediation**
  - (i) Phyto Extraction – Hyper accumulation
  - (ii) Enhanced Rhizosphere Phytoremediation
  - (iii) Phytostabilization
  - (iv) Phytodegradation
  - (v) Phytovolatilization

Greenhouse Gas (GHG) Emission and Mitigation

- Gases that create a greenhouse effects over the earth surface, thereby increasing the temperature of the earth.
- The principal GHGs are carbon dioxide (CO_2), Nitrous oxide (N_2O) and methane (CH_4). Other GHGs includes chlorofluorocarbons.

Causes of GHG Emission includes:

- Fossil fuel combustion.
- Land use changes i.e. deforestation, biomass burning.
- Agriculture sources of carbons (CO_2):
  - Operation of farm machinery.
  - Fertilizers and agriculture lime.
  - Pesticides.
  - Irrigation
  - Seed production

Impacts of GHG on global climate changes:

- Increased temperatures
- Severe droughts
- Floods
- Sea level rise

Mitigation

One major way of reducing GHGs emission is Soil Carbon Sequestration.

Carbon Sequestration is the process of carbons storage in soils. Consequently soils could act as carbon "sinks".

Carbon "sinks" are: Forest soils, grassland soils, ocean, & agricultural land.

Effects of Global Warming

The resultant effects of greenhouse gases emission is largely seen in global warming which causes:

- Weather – climate change
- Ice/glacier melt
- Sea level rise

- Agricultural productivity reduction
- Plants and Animals health effects
- Human health effects

WASTE DISPOSAL; EVALUATION AND MANAGEMENT OF CONTAMINATED LAND SITES

Waste is any plastics, paper, glass, metal, food, chemicals, wood, oil, soil, effluents, liquids that have been discarded. Waste gets generated from commercial, household and industrial sources. Sewage sludge is another source. Domestic & municipal waste is generated by the consumption of goods, manufacturing, sewage treatment, agriculture, the production & disposal of hazardous substance and construction. Wastes disposal has been a problem, and after industrialization the problem has only compounded. In the past, trash was carried to the outskirts of cities and discarded in the open, but now that can no longer be done. Over time, various wastes disposal methods have been devised, like composts, burning, landfill, biological reprocessing, etc.

Types of Wastes

There are basically three types of wastes generated and they are classified based on their chemical, biological & physical characteristics viz:

- **Solid wastes** include materials like mining wastes and industrial wastes besides household garbage.
- **Liquid wastes** are those in which the composition of solids is less than 1% and there is a high concentration of metals & salts.
- **Sludge** Contains a mixture of solid and water.

Waste Disposal is the management of wastes to prevent harm to the environment, injury or long term progressive damage to health. Disposal of waste is where the intention is to permanently store the wastes for the duration of its biological and chemical activity, such that it is rendered harmless. Disposal of wastes can be made to three locations namely surface water, atmosphere and land. The land represents not only the appropriate disposal medium for many wastes but also provides opportunity to manage wastes with minimum of adverse environmental effects.

Methods of Waste disposal

(a) Integrated wastes management (IWM)

Integrated wastes management using LCA (life cycle analysis) attempts to offer the best option for waste management. IWM entails using a combination of techniques and programmes to manage the municipal/urban wastes stream.

(b) Landfill

Disposing of wastes in a landfill involves burying the waste, and this remains a common practice in most countries. Landfills are often established in abandoned or unused quarries, mining voids or borrow pits. A properly designed and well-managed landfill can be a hygienic and relatively inexpensive methods of disposing off waste materials.

Advantages of Landfill	Disadvantages of landfill
Volume can increase with little addition of people/equipment	Completed landfill areas can settle and requires maintenance
Filled land can be refused for other community purposes	Requires proper planning, design, and operation

(c) Incineration

Incineration is a disposal method in which solid organic wastes are subjected to combustion so as to convert them into residue and gaseous products. This method is useful for disposal of residue of both solid waste management and solid residue from waste.water management.

Advantages of incineration	Disadvantages of incineration
• requires minimum land • can be operated in any weather • produces stable odour free residue • refuse volume is reduced by half	• expensive to build and operate • high energy requirement • requires skilled personnel and continuous maintenance • unsightly – smell, waste, vermin

(d) Recycling

Recycling refers to the collection and reuse of waste materials such as empty beverage containers. The materials from which the items are made can be reprocessed into new products. Material for recycling may be collected separately from general waste using dedicated bins and collection vehicles, or stored directly from mixed waste streams.

Advantages of recycling	Disadvantages of recycling
• key to providing a livable environment for the future	• expensive • some wastes cannot be recycled • technological push needed • separation of useful material from waste may be difficult

(e) Ocean Dumping

This is the deposition of waste into the water bodies, particularly the ocean. Controversy surround ocean dumping as a waste disposal method. Although the waste may provide nutrients for some sea life, it's widely believed that the harmful effects would outweigh any benefits.

Advantage of ocean dumping	Disadvantages of ocean dumping
• convenient • inexpensive • source of nutrients, shelter and breeding	• ocean overburdened • destruction of food sources • killing of plankton • desalination

(f) Open Dumping

This method involves disposing waste on a land. It is not without its own shortcoming.

Advantages of open dumping	Disadvantages of open dumping
• inexpensive	• health hazard - insects, rodents etc. • damage due to air pollution • ground water and run-off pollution

(g) Biological reprocessing

Waste materials that is organic in nature, such as plant material, food scraps, and paper products, can be recycled using biological composting and digestion processes to decompose the organic matter. The resulting organic material is then recycled as mulch or compost for agricultural or landscaping purposes.

(h) Energy recovery

The energy content of waste products can be harnessed directly by using them as a direct combustion fuel, or indirectly by processing them into another type of fuel. Recycling through thermal treatment ranges from using waste as a fuel source for cooking or heating to anaerobic digestion and the use of the gas fuel , to fuel for boilers to generate steam & electricity in a turbine.

(i) Avoidance and reduction methods

An important method of wastes management is the prevention of waste material being created, also known as waste reduction. Methods of avoidance includes reuse of second hand products, repairing broken items instead of buying new, designing products to be refillable or reusable (such as cotton instead of plastic shopping bags), encouraging consumers to avoid using disposable products (such as disposal cutlery), removing any food/liquid remains from cans, packing, and designing products that use less material to achieve the same purpose (for example, light weighting of beverage cans.)

Waste management

Waste management is the collection, transport, processing, recycling or disposal, and monitoring of waste materials. The term usually relates to materials produced by human activity, and is generally undertaken to reduces their effects on health, the environmental or aesthetics. Waste management is also carried out to recover resources from it.

There are a number of operations involved such as the collection, transport, processing, recycling or disposal, and monitoring of waste materials. The term usually relates to materials produced by human activity, and is generally undertaken to reduce their effects on health, the environment or aesthetics. Waste management is also carried out to recover resources from it.

- **Waste hierarchy-** The waste hierarchy refers to the "3 Rs" ie. reduce, reuse and recycle, which classify waste management strategies according to their desirability in terms of waste minimizations. The waste hierarchy remains the cornerstone of most waste minimization strategies. The aim of the waste hierarchy is to extract the maximum practical benefits from products and to generate the minimum amount of wastes.
- **Extend producer responsibility-** Extended Producer Responsibility (ERP) is a strategy designed to promote the integration of all costs associated with products throughout

their life cycle (including end –of- life disposal costs) into the market price of the product. Extended producer responsibility is meant to impose accountability over the entire lifecycle of products and packaging introduced to the market. This means that firms which manufacture, import and / or sell produce to be responsible for the products after their useful life as well as during manufacture.

- **Polluter pays principle-** This is a principle where the polluting party pays for the impacts caused to the environment. With respect to waste management, this generally refers to the requirement for a waste generator to pay for appropriate disposal of the waste.

3 CHAPTER Causes, Types, & Sources of Soil Pollution

Land pollution is contaminating the land surface of the earth through dumping of urban waste matter, dumping of industrial waste, mineral exploitation, and misusing the soil by harmful agricultural practices.

The soil gets polluted by the chemicals in pesticides and herbicides used for agricultural purposes along with waste matter being littered in urban areas such as roads, parks, & streets.

Land pollution results in the degradation of the Earth's land surface through misuse of the soil by poor agricultural practices, mineral exploitation, industrial waste dumping, and indiscriminate disposal of urban wastes, It includes visible waste and litter as well as pollution of the soil itself.

COMPOSITION OF LAND POLLUTION

Solid Waste

Solid wastes are unwanted solid materials such as garbage, paper, plastics & other synthetic materials, metals, & wood. Semisolid or solid matters that are created by human or animal activities, and which are disposed because they are hazardous or useless are known as solid waste. Most of the solid wastes, like paper, plastic containers, bottles, cans, and even used cars & electronic goods are not biodegradable, which means they do not get broken down through inorganic or organic processes. Thus, when they accumulate they pose a health threat to people, plus, decaying wastes also attract household pests and result in urban areas becoming unhealthy dirty, & unsightly places to reside in. Moreover, it also causes damage to terrestrial organisms, while also reducing the use of the land for other, more useful purposes.

Areas where wastes are buried, called landfills, are the cheapest and most common disposal method for solid wastes worldwide. But landfills quickly become overfilled any may contaminate air, soil, & water. Incineration, or burning, of waste reduces the volume of solid waste but produces dense ash wastes (some of which become airborne) that often contain dangerous concentrations of hazardous materials such as heavy metals and toxic compounds. Composting, using natural biological processes to speed up the decomposition of organic wastes, is an effective strategy for dealing with organic garbage and produces a material that can be used as a natural fertilizer. Recycling, extracting and reusing certain waste materials, has become an important part of municipal solid waste strategies in developed countries. Expanding recycling programs worldwide can help reduce solid waste pollution, but the key to solving severe solid waste problems lies in reducing the amount of waste generated. Waste prevention, or source reduction such as altering the way products are designed or manufactured to make them easier to reuse, reduces the high costs associated with environmental pollution.

Some of the sources of solid waste that cause land pollution are:

- **Wastes from Agriculture:** This comprises of waste matter produced by crop, animal manure, and farm residues.
- **Wastes from industries:** Piles of coal refuse and heaps of slag.
- **Wastes from Industries:** Industrial waste matter that can cause land pollution includes paints, chemicals, and so on.
- **Solids from Sewage Treatment:** Wastes that are left over after sewage has been treated biomass sludge, and settled solids.
- **Ashes:** The residual matter that remains after solid fuels are burned.
- **Garbage:** This comprises of waste matter from food that are decomposable and other waste matter that are not decomposable such as glass, metal, cloth, plastic, wood, paper, and so on.

Soil Pollution

Soil pollution is the buildup in soils of persistent toxic-compounds, chemicals, salts, radioactive materials, or disease-causing agents, which have adverse effects on plant growth and animal health. As of now, soil pollution is not widespread. Although the application of fertilizers containing the primary nutrients, nitrogen, phosphorus, & potassium, has not led to soil pollution, the application of trace elements has. The irrigation of arid lands often leads to pollution with salts. Sulfur from industrial wastes has polluted soils in the past, as has the accumulation of arsenic compounds in soils following years of spraying crops with lead arsenate. The application of pesticides has also led to short-term soil pollution.

Unhealthy soil management methods have seriously degraded soil quality, caused soil pollution, and enhanced erosion. Treating the soil with chemical fertilizers, pesticides and fungicides interferes with the natural processes occurring within the soil and destroys useful organisms such as bacteria fungi, & other microorganisms. Soil pollution is chiefly caused by chemicals in pesticides, such as poisons that are used to kill agricultural pests like insects & poisons and herbicides that are used to get rid of weeds.

Hence, soil pollution results from:

(a) unhealthy methods of soil management.
(b) harmful practices of irrigation methods.

Causes of Land Pollution

The following are the major causes of land pollution:

- **Increase in urbanization:** Construction uses up forest land. More constructions mean increase in demand for raw materials like timber. This leads to the exploitation and destruction of forests. There is more demand for water. Reservoirs are built leading to the loss of land.
- **Increase in solid waste generation:** Every single day, tons and tons of domestic waste are dumped ranging from huge pieces of rubbish such as unused refrigerator to fish bones. If all these wastes are not disposed off properly, the damage they can do to the environment and human kind is enormous. We throw away more things today and this causes an increase in the quantity of solid waste. This has given rise to new dumping grounds being sought after.

- **Agricultural activities**: Besides domestic waste, pesticides & herbicides are used by farmers to increase crop yield. During open cast mining, huge holes are dug in the ground and these form dangerously deep mining pools. Heaps of mining waste are left behind and these waste often contain several poisonous substances that will contaminate the soil.

  Pollution sources include plastic factories, chemical plants, oil refineries, nuclear waste disposal activity, large animal farms, coal-fired power plants, metals production factories and other heavy industries.

Causes of Land Pollution can be summarized as follows.

- **Degenerative Actions:** encompass a lot of human actions, including deforestation, overuse of pesticides & chemical fertilizers, mining, inefficient and / or inadequate waste treatment, landfill, litter, etc. Many of these are unavoidable; however, definitely the severity can be reduced by adequate corrective measures that needs to be taken.
- **Misuse of Land**: It mainly refers to felling of trees to clear land for agriculture, as well as processes like desertification and land conversion. Desertification is when anthropogenic effects of human development and / or other actions converts a piece of (essentially) fertile land into desert land or dryland. Land once converted to desert land can never be reclaimed by any amount of corrective measures. This is also a serious issue because it not only affects the land, but also the overall biodiversity of a place, specially when land is cleared for agriculture. A lot of indigenous flora and fauna is lost in the process.
- **Inefficient Use of Land**: It is an ecosystem in which millions of living creatures live and interact. Soils provide many important functions for plants, animals and humans. Soil is extremely important as a filter for removing pollution from our drinking water and helping to regulate the flow of water through the landscape. Most rainwater ends up moving into the soil before it gets to plant roots, the aquifers or the river. Soils are important in the story of climate change, since it is capable of acting both as a source and sink of carbon. Soils also help in regulating other greenhouse gases such as nitrous oxide and methane.

  Few uses of soil are: we grow our crops in soil; forest grows on soil and soil filters the water we drink.

Causes of Soil Pollution

- **Discharge of industrial waste (solid, liquid & gases) into the soil:** Industry often produces huge quantities of unwanted chemicals, which, if allowed to come in contact with the soil, can cause extensive pollution. If cleaning agents such as detergents are released irresponsibly, they can cause pollution to the soil and again to the drinking water supplies.
- **Percolation of contaminated water into the soil**
- **Excess application of pesticides, herbicides or fertilizer in agricultural fields:** The main problem on modern intensive farms is the increased use of and dependence on chemical fertilizers, pesticides & insecticides. If applied in the wrong strengths, these can remain in the soil, and will also leach off the fields & into local water supplies. These toxic chemicals are taken up by the plants & animals, that we eat and so enter in our bodies. This can lead to cancers and wide variety of other illnesses.

- **Dumping of solid waste (Municipal solid waste, Bio-medical waste, e-waste etc) and from a landfill:** We produce vast quantities of varieties of waste every year, much of them could and should be recycled or allowed to degrade naturally. Much of our waste is either incinerated with potential problems of air pollution or buried in landfill sites, where leaching of waste builds up problems for the future.
- **Deforestation and Soil erosion:** Deforestation means the conversion of forest land for agricultural uses to produce food & materials and for other development uses. When forests are cleared (vegetation is removed), soil is exposed to heat and rainwater. Deforestation causes increased losses of carbon, nitrogen, phosphorus, and sulphur from terrestrial ecosystems. Nitrogen & Sulphur are lost by the same pathways, and additionally by leaching of nitrate & sulphate to stream-water & ground-water and by the anaerobic production of N & S . Soil erosion is a process that involves the removal and transport of soil by wind & water. It is a natural process but human activities such as changes in farming & land management practices can cause it to occur much faster than under natural conditions. The rapid growth of population has resulted in increased cultivation of land. This puts more pressure on land and leads to soil losing its structure & cohesion, which means that it can be more easily degraded.
- **Rupture of underground storage tanks:** Underground storage tanks can be a major source of pollution, if they leak.
- **Acid rain:** The acidic water falls on the soil and pollutes it by making the soil acidic. Soil loses its nutrient due to high solubility in acidic water. Soil polluted by acid rain has an impact on plants also. It disrupts the soil chemistry, reduces plant's ability to take up nutrients and undergoes photosynthesis.
- **Overburdens of mines:** Huge holes are dug in the ground during mining . Waste is obtained from debris left on site in the form of spoil heaps. This waste contains toxic and poisonous substances which then leach into the soil due to rainfall.
- **Sewage Sludge:** Sludge originates from the process of treatment of waste water. Due to the physical-chemical processes involved in the treatment, the sludge tends to concentrate heavy metals and poorly biodegradable trace organic compounds as well as potentially pathogenic organisms (viruses, bacteria etc) present in waste waters. Sewage sludge contain organic waste material, nitrogen, phosphorus, traces of many pollutants used in our modern society, pathogenic bacteria, viruses and protozoa along with other parasitic helmets, which can give rise to potential hazards to the health of humans, animals & plants when applied on soil.

Causes of Soil Pollution can be summarized as follows:

- Increasing urbanization
- Disposal of untreated wastes
- Indiscriminate use of Agrochemicals & Agricultural practices
- Unscientific mining
- Dumping Industrial Wastes
- Accidental Pollution/Leakages
- Outdated technology, inadequate treatment & safety management of chemicals and waste
- Lack of Designed Engineered Landfills

Types of Land Pollution

Land pollution can be caused due to various reasons. Each of these types have a different adverse kind of effect on the soil. Some of the effects are so adverse that they can cause the soil to be damaged permanently.

The degradation of the land surface of the earth caused due to different human activities is land pollution. The misuse of land resources is also categorized under land pollution. It can also pose danger to human life. Basically, pollution started after industrialization. Till goods were manufactured only by hands, the earth was not abused. However, with industrialization started the need to produce more in the shortest period of time lead to misuse of the land resources as well. There was also a sudden need to increase agriculture production, which in turn caused further pollution. To industrialization, there was addition of urbanization, which led to pollution of land in the industrial area. In other words, there are many ways due to which land can be polluted, which are causing destruction of the surface of the earth.

DIFFERENT TYPES OF LAND POLLUTION

Agricultural Land Pollution

To increases the yield of products, more and more fertilizers are used, along with large quantities of pesticides, insecticides & fungicides. They are all rich in chemicals, which are able to kill the pests and increase production, but at the cost of stripping the soil of its minerals & other useful compounds. Along with pests, they cause harm to birds, mammals, fish and beneficial microbes as well. Some of The herbicides used to get rid of weeds are so harmful, that they can prove to be fatal even when they are used in small quantities.

Chemical Land Pollution

Waste material produced from different industries contributes to land pollution. More often than not the waste is not treated to get rid of toxicity, which cause damage to the soil, Responsible methods have to be used before the chemicals are disposed off.

Garbage

There is a lot of garbage that is generated in each of our houses, as well as at places of work. Garbage can be in form of metal, glass, plastic, wood or even paper. This garbage finds its way to the dumping grounds, which in turn give rise to landfills. Some of the garbage is biodegradable product. A number of products manufactured using plastic make piles of waste and it becomes very difficult to get rid of them. Hence, it is important that one turns them to reusable products, which will help in reducing garbage.

Refuse

Food that has gone bad or is not consumed also makes for a large chunk of land pollution. If the restaurant waste has to be analyzed, one will see that the food wasted would have been able to feed a number of hungry people. Reducing the quantity of refuse generated will help in reducing land pollution.

Littering

Littering is a problem in all parts of the world. The effect of littering is that many animals choke on these littering pieces thrown all around. If there is food in the packets and it has gone bad the animals may be poisoned with the food which can lead to their death. Piles of litter is gathered from different parts, but the next question remains is what should one do with it, as it will take years to get rid of some and there is some, which will continue to occupy space around. Some of the items are so high in chemicals, that even if they alter over a period to time, they will alter the soil composition.

These were some of the ways in which land is polluted. Prevention of land pollution is the need of the hour. If stringent measures are not taken now, the end of the world is not very far. All of these causes a number of problems like birth defects, mental dysfunction, impairment of the reproductive organs, genetic disorders, etc.

Types of Soil Pollution can be summarized as follows:

(a) Industrial Wastes Leading to Soil Pollution

Sources and effects: Pulp and paper mills, chemical & fertilizer industries, oil refineries, coal & mineral mining, sugar factories. These pollutants affect and alter the chemical & biological properties of soil. It disturbs human food chain and has serious effects on human life.

(b) Urban waste Leading to Soil Pollution

Sources and effects: Plastic, Glasses, metallic cans, papers, rubbers, street garbage & rubbish materials, sweeping and other discarded manufactured products. These are also dangerous for human beings; Non-degradable wastes harm the environment.

(c) Agriculture Practices Leading to Soil Pollution

Sources and effects: Huge quantities of fertilizers, pesticides, herbicides, and weedicides are added to increase the crop yield. Apart from these farm waters, manure, slurry etc also lead to soil pollution. Other sources include eroded soil containing inorganic chemicals, & modern agricultural methods.

(d) Radioactive/Nuclear Pollutants Leading to Soil Pollution

Sources and effects: These result from explosion of nuclear bombs and operation of nuclear reactors. Radioactive wastes penetrate the soil and accumulate there causing land pollution. Examples include radio nuclides of Th, U, K & C and nuclear reactor wastes like R_u, I, B_a & C_s which emits gamma radiation. Rain water carries the harmful wastes into the soil.

(e) Biological Agents Leading to Soil Pollution

Sources and effects: Soil gets large quantities of human, animal and birds excreta which constitute the major source of land pollution by biological agents.

Land pollution is caused when wastes from industrial factories, agricultural wastes and the contributions of humankind amass to what we see and hear about today. It not only diffuses harmful chemicals & toxins into the atmosphere, but settles into the soil and creeps into water sources, affecting life that thrives under these. Not only are people's health suffering detriment due to pollution, but plants and animals are taking a hit as well.

Besides air & water pollution, land pollution is an everyday problem that is on the rise and harming everything that comes, in its way. The most virulent contributing factor when it comes to pollution is automobile emissions and wastes that industries release on a daily basis.

The dangerous factor that contribute to land pollution are:

- Oil refineries
- Debris from construction sites
- Littering
- Sewage of human waste
- Agricultural toxic waste and chemicals
- Mining
- Demolition
- Nuclear and chemical plants
- Deforestation
- Household wastes
- Landfills that are overflowing
- Antifreeze/oil leaking from cars

The major sources of this kind of pollution are those that cannot be dealt with, since industries will continue to run and automobiles will still be bought while increasing in number by the day. There's nothing we can do about what still has to run its course, and if people have to saturate the land with harmful elements that do more damage than it ever could in the past, there's nothing we can do to put a complete stop to it. So for us to play a small part when it comes to damage control in our neighborhoods and in our lives, we can practice some ways on controlling what we see when it comes to destroying nature and harming humankind.

- Use paper bags or recyclable ones instead of plastic. Carrying own bag would help instead of taking these from stores that distribute plastic.
- Eat foods that are organic in nature to ensure that you are ingesting only foods that have not been subjected to pesticides or chemical based fertilizers.
- Biodegradable products should be a priority on your list of to-buy items.
- Avoid dumping motor oil on the ground since this can cause the soil to absorb it, affecting plant and animal life.
- Avoid spilling engine oil onto the ground as well.
- Wastes and chemicals should be stored in spill containers.
- Avoid littering even if the urge to do so arises. Wait until you spot a trash can and then dispose of what you need to.
- Set an example to those who openly contaminate the area by dumping food/drink packages outdoors.
- Encourage others to plant trees as part of a project to help the environment stay as green as possible.

Categories of Sources of Land Pollution are as follows:

The sources of land pollution can be categorized as under:

- Agriculture
- Quarrying and mining

- Sewage sludge
- Dredged spoils
- Household domestic solid waste
- Demolitions and constructions wastes
- Industrial activities

Sources of Land Pollution

Land Pollution mainly results from the following sources:

- Industrial wastes
- Urban or commercial wastes,
- Agricultural wastes;
- Municipal wastes;
- Chemical and metallic pollutants:
- Biological reagents.

Hazardous wastes or Radioactive pollutants and Mining wastes

Pollution through Industrial Wastes

The rapid growth of industries has resulted in the release of lot of industrial waste on the land surface. The quantity of these wastes depends on the type of raw materials and chemicals used. It has been estimated that nearly 50% of the raw materials ultimately become waste products in the industries and about 20% of these wastes are extremely harmful.

The industrial wastes are mainly discharged from pulp & paper mills, chemical industries, oil refineries, sugar factories, tanneries, textile mills, steel plants, distilleries, fertilizer plants, pesticide industries, coal & mineral mining industries, metal processing industries, drugs, glass, petroleum and engineering industries etc.

Besides, thermal, electric power plants generate fly ash (un-burnt brownish black substance) which severely pollute air, water and land. The industrial wastes may be either organic or inorganic in nature. Also, these may be bio-degradable or non-bio-degradable.

The industrial pollutants affect and alter the chemical & biological characteristics of soil. The pollutants from soil enter into food chain, disturb the biochemical processes and finally induce serious hazards to living organisms.

Mahim Bay in Mumbai is chronically polluted due to addition of million tons of industrial wastes & of domestic sewage every year causing a serious threat to aquatic environment.

Pollution through Urban or Commercial Wastes

Urban wastes include both commercial and domestic wastes consisting of dried sludge of sewage. The solid urban waste is commonly known as refuse. The refuse contains garbage and rubbish materials such as plastics, glasses, metallic cans, fibers, rubbles, street sweepings, fuel residues, containers, abandoned vehicles and other discarded manufactured products.

Urban wastes are even more dangerous than the industrial waste because of their non-biodegradable nature. The leachates from dumping sites and disposal tanks of sewage mixed with industrial effluents & wastes become extremely toxic to living organisms. The leachates coming out of polluted soil contain partly decomposed organic material especially food remnants, vegetables, toxic hydrocarbons and pathogenic bacteria.

The maximum contribution to land pollution in India comes from urban areas, specifically from unplanned industrial progress in and around these areas. It is estimated that around 20 million tonnes of solid waste is dumped on land surfaces in India alone causing chronic pollution.

Pollution through agricultural wastes

Agricultural wastes include roots & stems of the crops, straw, hay, dung, food articles etc. These are also known as farm wastes. In addition to farm wastes, use of excessive fertilizers, pesticides, herbicides, weedicides and soil conditioning agents, induces land pollution to a greater extent. Although farm wastes are used and reused in a number of ways in developing countries, it has become a problem in developed countries.

When the unused fertilizers are contaminated with other synthetic organic pollutants, the water present in the soil gets polluted. Although fertilizers are retained by the crops & soils, there are reports that if nitrate & phosphate concentration exceeds one part & thirty part per hundred per million parts of water respectively, there happens a number of health hazards and eutrophication, chocking the whole stretch of aquatic ecosystem.

The natural environment, particularly the land, is also under a stress due to different kinds of pesticides used to control pests. Mainly the chlorinated hydrocarbons (like DDT, BHC, aldrin, endrin, dieldrin etc.) and organo phosphates (like malathion, parathion, ethion, fenthion etc.) have tremendous effect on land pollution. The remnant of these pesticides get absorbed by soil particles and coexist within biological system for a longer period of time.

In addition to fertilizers, pesticides & biocides, soil conditioners & fumigants (which contain several toxic metals like As, Cd, Hg, Co, etc.) are also added to the land system for increasing soil fertility and killing hazardous insects. These chemicals having long lifetime, accumulate on the soil and then enter the food chain through the growing crops, causing a number of health hazards.

Pollution through municipal wastes

The municipal wastes include a number of solid wastes like papers, plastics, metal cans, glass bottles, aluminum foils, metal junks, garbage's mostly coming domestic uses etc. In addition, these also contain organic matters, compostable matters, ash stone, coal, straw, leaves, rotten vegetables, polythene bags, human excreta, animal wastes etc.

In developed countries, these are dumped either in underground pits or in sea. But in developing countries like India, the municipal wastes are kept in certain dump land sites. When the municipal solid wastes are kept for a longer period of time these pollute the environment by foul smell and releasing a number of harmful gases.

The municipal waste substances are also classified as:

- Non-combustible wastes;
- Highly combustible wastes,
- Combustible wastes;

Pollution through chemical and Metallic Pollutants

The industries like textiles, paints dyes, soap & synthetic detergents, tanneries, drugs, batteries, cement, asbestos, rubber, petroleum, paper & pulp, sugar, steel, electroplating and metal industries release harmful effluents in soil & water hazards in living world, It is reported that synthetic chemicals and fertilizers are the source of trace metals such as As, Pb, V, Mn, Hg, Mo, Si, etc, which are added to the soil in one or other from.

When soil is found to be complexed with organic chemicals and toxics or its ions, these affect the soil texture, its fertility and stabilization of organic matter. Excess of sulphur in soil may be absorbed by plant leaves as SO_2 injuring plant tissues.

The presence of high concentrati8on of Mg, Fe, Zn and V, causes lethal effects on crop production. The metallic contaminants in soil are considered to be the indestructible poisons and their accumulation in plants & water may be highly dangerous because these inhibit the production of atmospheric oxygen and reduce the rate of photosynthesis. Digested sewage sludge & heavy application of manure to soil, causes chronic salt hazards to plants.

Pollution by Biological Agents

The major source of land pollution by biological agents comes from the excreta of human beings, animals and birds. The sewage sludges contain a number of live viruses & viable intestinal worms which take shelter on lands and cause a number of diseases by attacking the living organisms.

The pathogenic organisms that pollute the soil may be divided into the following classes:

(a) Pathogenic organisms occurring naturally in contaminated soil:

The soil is residence of a number of bacteria, fungi, algae, protozoans, actiomycetes, nematodes, rotifiers etc. These modulate soil fertility and soil texture. Besides, these attack the different part of the plant, specially the roots, causing significant damage.

(b) Pathogenic organisms excreted by man:

Human excreta contain pathogens such as enteric bacteria and parasitic worms, which reside in soil. These organisms enter into human body either via chain or by direct contact with contaminated soil. The sanitary habits of people help in the repetition of the cycle of infection with soil transmitted pathogens from man to soil and from soil to man.

(c) Pathogenic organisms excreted by animals: This class includes pathogenic bacteria and parasitic worms excreted from animals. These organisms are transmitted from animals to soil and then soil to man.

Hazardous Wastes (Radioactive Wastes):

In addition to chemical reagents, radioactive wastes cause serious environmental hazards. Radioactive substances produced from nuclear explosion, nuclear reaction & from nuclear testing laboratories, enters into the soil and accumulate there causing land pollution.

Radioactive waste contains several radio nuclides such as Ceasium-137, Strontium-90, Iodine-129 and isotopes of radioactive elements. These wastes are highly dangerous and causes cancer in living organisms & also degrade genetic materials like DNA & RNA inducing mutation. Nuclear reactors produce wastes containing Ruthenium-106, Iodine-131, Barium0140 Lanthanum-140 etc.

Examples of radioactive waste includes

- Radionuclides of Radium, Thorium, Uranium, isotopes of Potassium (K-40) and Carbon (C-14) are commonly found in soil, rock, water, air etc.
- Explosion of hydrogen weapons and cosmic radiations include neutron, proton reactions by which Nitrogen (N-15) produces C-14. This C-14 participates in Carbon metabolism of plants which is then transmitted into animals and human beings.
- Radioactive waste contains several radio nuclides such as Strontium-90, Iodine-140, Cesium-137 and isotopes of iron which are most injurious. Strontium get deposited in bones and tissues instead of calcium.
- Nuclear reactors produce waste containing Ruthenium-106, Iodine-131, Barium-140, Cesium-144 and Lanthanum-140 along with primary nuclides Sr-90 with a half life 28 years and Cs-137 with a half life 30 years. Rain water carries Sr-90 and Cs-137 to be electrostatic forces. All the radio nuclides deposited on the soil emit gamma radiations.
- Biological agents: Soil gets a large amount of human, animal and bird excreta which constitute a major source of land pollution by biological agents.

  Heavy application of manures and digested sludge can cause serious damage to plants within a few years.

Pollution through Mineral Wastes

During mining operations, huge quantities of wastes are dumped on land surface. For example, during mining of metal ores, huge quantities of earthen materials, rocks and other wastes are created. Similarly, during coal mining surficial materials are removed to expose coal seam. Thus huge quantity of waste mining materials is created which not only occupy valuable land surface but also create nuisance due to the presence of toxic chemicals.

Sources & Mechanisms of Land Pollution is tabulated below:

Table 3.1: Sources & Mechanisms of Land Pollution

S. N	Source	Reason/Cause/Mechanism
1	Agriculture	(i) Accumulation of animal manure. (ii) Excessive input of chemical fertilizers (iii) Farm residues. (iv) Illicit dumping of tainted or harmful crops on land.
2	Quarrying and Mining	(i) Use of explosives to blow up mines. (ii) Mining tailings and refuse. (iii) discharge of acidic effluents. (iv) Use of heavy machineries and large number of vehicles.
3	Sewage sludge	(i) Untreated municipal sludge.
4	Dredged spoils	(i) Improper method of dredging disposals.

5	House hold or Domestic solid waste.	(i) Improper and inefficient method of collection, transportation and disposal of solid waste causes accumulation. (ii) Improper sanitation system.
6	Demolition & Construction wastes	(i) Non-biodegradable rubbles or debris create unsightly condition.
7	Industrial wastes	(i) Discharge of untreated effluents on the land for disposal. (ii) Deposition of process and effluent treatment sludge. (iii) release of gaseous pollutants such as SO_2 causing acid rains.

APPENDIX 3.1

Causes & effects of soil pollution

Soils are formed by the decomposition of rock and organic matter over many years. Soil properties vary from place to place with differences in bedrock composition, climate and other factors. At times, the amounts of some soil elements and other substances may exceed levels recommended for the health of humans, animals, or plants. Certain chemical elements occur naturally in soils as components of minerals, yet may be toxic at some concentration. Other potentially harmful substances may end up in soils through human activities.

Soil properties are affected by past land use, current activities on the site, and nearness to pollution sources. Human activities have intentionally added substances such as pesticides, fertilizers and other amendments to soils. Accidental spills and leaks of chemicals used for commercial or industrial purposes have also been sources of contamination. Some contaminants are moved through the air and deposited as dust or by precipitation.

Fate of Contaminants in soil is shown below in figure.

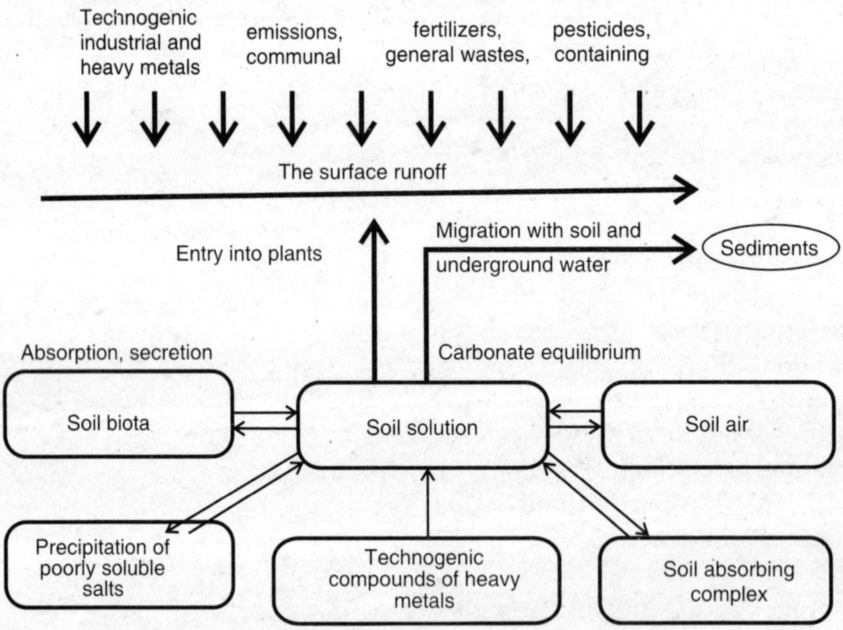

Fig.3.1F_1: Fate of Contaminants in Soil

Once contaminants are in soils, where they go and how quickly they travel depends on many factors. Some organic (carbon-based) contaminants can undergo chemical changes or degrade into products that may be more or less toxic than the original compound. Chemical elements (such as metals) cannot break down, but their characteristics may change so that they can be more or less easily taken up by plants or animals. Different contaminants vary in their tendency to:

- End up in water, help in the soil or in the underlying groundwater (by leaching through the soil),
- Volatilize (evaporate) into the air, or
- Bind tightly to the soil.

  The characteristics of the soil also affect the fate of contaminants and whether they can be readily taken up by plants or animals. Site management and land use (such as gardening practices) can affect some soil characteristics. Important soil characteristics include:
- Soil mineralogy and clay content (soil texture);
- pH (acidity) of the soil;
- Amount of organic matter in the soil;
- Moisture levels;
- Temperature; and
- Presence of other chemicals.

Soil is fundamental to human life on Earth. Most plants require a soil substrate to provide water & nutrients, whether we farm the plants directly or consume animals that feed on the plants, this means that we don't eat without soil.

Soil is a vital part of the natural environment. It is just as important as plants, animals, rocks, landforms and rivers. It influences the distribution of plant species and provides a habitat for a wide range of organisms. It controls the flow of water & chemical substances between the atmosphere & the earth, and acts as both a source & store for gases (like oxygen & carbon dioxide) in the atmosphere.

Soil, together with the plant & animal life it supports, the rock on which it develops its position in the landscape and the climate it experiences, form an intricate natural system soil is always responding to change in environmental factor, along with the influences of man & land use. Some changes in the soil can be of short duration and reversible, others will be a permanent feature of soil.

Soil pollution is the reduction in the productivity of soil due to the presence of soil pollutants. Soil pollutants have an adverse effect on the physical chemical & biological properties of the soil and reduce its productivity. Pesticides, fertilizers, organic manure, chemicals, radioactive wastes, discarded food, clothes, leather goods, plastics, paper, bottles, tin cans & carcasses-all contribute towards causing soil pollution. Chemicals like iron, lead, mercury, copper, zinc cadmium, aluminium, cyanide, acids & alkalies etc. are present in industrial wastes and reach the soil either directly with water or indirectly through air, (e.g. through acid rain).

The improper and continuous use of herbicides, pesticides & fungicides to protect the crops from pests, fungi etc. alter the basic composition of the soils and make the soil toxic for plant growth.

They accumulate in the soil as they degrade very slowly by soil and water bacteria. Consequently, they have a very deleterious effect on the plant growth; stunting their growth and reducing the yield & size of fruit. Their degradation products may be absorbed by the plants from where they reach the animals and man through the food chains.

Radioactive wastes from mining and nuclear processes may reach the soil via water or a 'fall-out'. From the soil they reach the plants and then into the grazing animals (livestock) from where ultimately reach man through milk & meat etc. resulting in retarded & abnormal growth of man. Human & animal excreta used as organic manure to promote crop yield, pollute the soil by contaminating the soil & vegetable crops with the pathogens that may be present in the excreta.

Nitrification, which is the process of forming soluble nitrates from the elemental atmospheric nitrogen or from originally harmless organic materials actually contribute towards water pollution when the nitrates leach out of the soil and accumulate to toxic levels in the water supply.

Therefore, intensification of agricultural production by practices of irrigation (causes salination), excessive fertilizers, pesticides, insecticides etc. have created the problems of soil pollution. Soil pollution can be checked by restricting the use of above mentioned soil pollutants, resorting to organic farming, adopting better agriculture practices etc.

Soil pollution is also caused by means other than the direct addition of xenobiotic (man-made) chemicals such as agricultural runoff waters, industrial waste materials, acidic precipitates, and radioactive fallout. Both organic and inorganic contaminants are important in soil. The most prominent chemical group of organic contaminants are important in soil. The most prominent chemical group of organic contaminants are fuel hydrocarbons, polynuclear aromatic hydrocarbons (PAHs), polychlorinated biphenyls (PCBs), chlorinated aromatic compounds, detergents, and pesticides. Inorganic contaminants species include nitrates, phosphates, and heavy metals such as cadmium, chromium & lead; inorganic acids; and radionuclides (radioactive substances).

Soil pollution can lead to water pollution, if toxic chemicals leach into groundwater, or if contaminated runoff reaches streams, lakes, or oceans. Soil also naturally contributes to air pollution by releasing volatile compounds into the atmosphere. Nitrogen escapes through ammonia volatilization and denitrification. The decomposition of organic materials in soil can release sulfur dioxide and other sulfur compounds, causing acid rain. Heavy metals and other potentially toxic elements are the most serious soil pollutants in sewage. Sewage sludge contains heavy metals and, if applied repeatedly or in large amounts, the treated soil may accumulate heavy metals and consequently become unable to even support plant life.

In addition, chemicals that are not water soluble contaminate plants that grow on polluted soils, and they also tend to accumulate increasingly toward the top of the food chain. The banning of the pesticide DDT in many countries resulted from its tendency to become more and more concentrated as in birds & their eggs. This occurs as creatures higher on the food chain ingest animals that were already contaminated with the pesticide from eating plants and other lower animals. Lake Michigan, as an example, has 2 parts per trillion (ppt) of DDT in the water, 14 parts per billion (ppb) in the bottom mud, 410 ppb in amphipods (tiny water flies and similar creatures), 3 to 6 parts per million (ppm) in fish such as Coho salmon & lake trout, and as much as 99 ppm ion herring gulls at the top of the food chain.

The rapid industrialization of agriculture, expansion of agriculture, expansion of the chemical industry, and the need to generate cheap forms of energy has caused the continuous release of man-made organic chemicals into natural ecosystems. Consequently, the atmosphere, bodies of water, and many soil environments have become polluted by a large variety of toxic compounds. Many of these compounds at high concentrations or following prolonged exposure have the potential to produce adverse effects in human and other organisms. These include the danger of acute toxicity, mutagenesis (genetic changes), carcinogenesis, and teratogenesis (birth defects) for humans & other organisms. Some of these man-made toxic compounds are also resistant to physical, chemical, or biological degradation and thus represent an environmental burden of considerable magnitude.

Numerous attempts are being made to decontaminate polluted soils, including an array of both in situ (on-site, in the soil) and off-site (removal of contaminated soil for treatment) techniques, none of these is ideal for remediating contaminated soils, and often, more than one of the techniques may be necessary to optimize the cleanup effort.

The most common decontamination method for polluted soils is to remove the soil and deposit it in landfills or to incinerate it. These methods, however, often exchange one problem for another: landfilling merely confines the polluted soil while doing little to decontaminate it, and incineration removes toxic organic chemicals from the soil, but subsequently releases them into the air, in the process causing air pollution.

For the removal and recovery of heavy metals various soil washing techniques have been developed including physical methods, such as attrition scrubbing & wet-screening, and chemical methods consisting of treatments with organic & inorganic acids, bases, salts and chelating agents. For example, chemicals used to extract radionuclides & toxic metals includes hydrochloric, nitric, phosphoric & citric acids, sodium carbonate & sodium hydroxide and the chelating agents EDTA & DTPA. The problem with these methods, however, is again that they generate secondary waste products that may require additional hazardous waste treatments.

In situ methods are used directly at the contamination site. In this case, soil does not need to be excavated, and therefore the chance of causing further environmental harm is minimized. In situ biodegradation involves the enhancement of naturally occurring microorganisms by artificially stimulating their numbers and activity. The microorganisms then assist in degrading the soil contaminants. Biodegradation is facilitated by aerobic soil conditions and soil pH in the neutral range (between pH 5.5 to 8.0), with an optimum reading occurring at approximately pH 7, and a temperature in the range of 20 to 30°C. These physical parameters can be influenced, thereby promoting the microorganism's ability to degrade chemical contaminants. Of all the decontamination methods bioremediation appears to be the least damaging and most environmentally acceptable technique.

The upper layer of the unsaturated zone of earth is mineral and organic constituents. Soil offers shelter, habitat for numerous organisms and is the living medium for plants. Soil is the basis for agriculture. All vegetation for human food and animal feed depend upon soil.

Enormous quantities of waste from man-made products are being released into the soil causing soil pollution. Polluted water also causes soil pollution. Soil pollution is caused due to unhygienic habits, agricultural practices and inappropriate methods of disposal of solid & liquid wastes. Soil pollution is also caused as a result of atmospheric pollution.

In industrialized countries, soil pollution is a result of use of chemicals in agriculture, dumping of waste materials, mining, smelting of metals and also dumping of domestic refuse & solids, untreated sewage and industrial wastes.

TYPES OF SOIL POLLUTION

Types of soil pollution can be distinguished by the source of the contaminant and its effects on the ecosystem. Types of soil pollution may be agricultural pollution, industrial wastes and urban activities.

Agricultural Pollution

- Agricultural processes contribute to soil pollution.
- Fertilizers increase crop yield but also cause pollution that impacts soil quality.
- Pesticides also harm plants and animals by contaminating the soil.
- These chemicals get deep inside the soil and poison the ground water system.
- Runoff of these chemicals by rain and irrigation also contaminate the local water system and is deposited at other locations.

Industrial Wastes

- About 90% of oil pollution is caused by industrial waste products.
- Improper disposal of waste contaminates the soil with harmful chemicals.
- These pollutants affect plant & animal species and local water supplies & drinking water.
- Toxic fumes from the regulated landfills contain chemicals that can fall back to the earth in the form of acid rain and can damage the soil profile.

Urban Activities

- Human activities can lead to soil pollution directly and indirectly.
- Improper drainage and increased run-off contaminate the nearby land areas or streams.
- Improper disposal of trash breaks down into the soil and it deposits a number of chemical & pollutants into the soil. These may again seep into groundwater or wash away in local water system.
- Excess waste deposition increases the presence of bacteria in the soil.
- Decomposition by bacteria generates methane gas contributing to global warming and poor air quality. It also creates foul odors and can impact quality of life.

Causes of Soil Pollution

Soil Pollution is a result of many activities by mankind which contaminate the soil. Soil pollution is associated with indiscriminate use of farming chemicals, such as pesticides fertilizers, etc. Pesticides applied to plants can also leak into the ground, leaving long-lasting effects. In turn, some of the harmful chemicals found in the fertilizers (e.g. cadmium) may accumulate above their toxic levels leading to the poisoning of crops. Heavy metals can enter the soil through the use of polluted water in watering crops, or through the use of mineral fertilizers. Faulty landfills, bursting of underground bins and seepage from faulty sewage systems could cause the leakage of toxins into the surrounding soil. Acid rain falls on the land, and could dissolve away some of the important nutrients found in soil, as such change the structure of the soil. Industrial wastes are one of the biggest soil-pollution factors. Iron & Steel, power and chemical manufacturing plants which use the earth as dumping ground often leave behind lasting effects for years to come. Fuel leakages from automobiles, which get washed by rain, can seep into the nearby soil, thereby polluting it. Deforestation is a major cause for soil erosion, where soil particles are dislodged and carried away by water or wind. As a result, the soil loses its structure as well as important nutrients found in the soil. Some the causes of soil pollution are as follows:

- Industrial effluents like harmful gases and chemicals.
- Use of chemicals in agriculture like pesticides, fertilizers and insecticides.
- Improper or ineffective soil management system.
- Unfavorable irrigation practices.
- Improper management and maintenance of septic system.
- Sanitary waste leakage.
- Toxic fumes from industries get mixed with rains causing acid rains.
- Leakages of fuel from automobiles are washed off due to rains and are deposited in the nearby soil.
- Unhealthy waste management techniques release sewage into dumping grounds and nearby water bodies.
- Use of pesticides in agriculture retains chemicals in the environment for a long time. These chemicals also effect beneficial organisms like earthworm in the soil leading to poor soil quality.
- Absence of proper garbage disposal system leads to scattered garbage in the soil. These contaminants can block passage of water into the soil and affects its water holding capacity.
- Unscientific disposal of nuclear waste contaminate soil and can cause mutations.
- Night soil contamination due to improper sanitary system in villages can cause harmful diseases.

Effects of soil pollution

Soil contamination leads to health risks due to direct and indirect contact with contaminated soil. Soil pollution causes huge disturbances in the ecological balance and the health of living beings on earth. Normally crops cannot grow and flourish in a polluted soil. However, if some crops manage to grow, then these crops might have absorbed the toxic chemicals in the soil and might cause serious health problems in people consuming them. Sometimes the soil pollution is in the form of increased salinity of the soil. In such a case. the soil becomes unhealthy for vegetation, and often becomes useless and barren. Then soil deaths of many beneficial soil organisms (e.g. earthworms) in the soil could take place other than further reducing the ability of the soil to support life. This occurrence could also have an effect on the larger predators (e.g. birds) and force them to move to other places, in the search of food. People living near polluted land tend to have higher incidences of migraines, nausea, fatigue, skin disorders and even miscarriages. Depending on the pollutants present in the soil, some of the longer-term effects of soil pollution include cancer, leukemia, reproductive disorders, kidney system disorder, and central nervous system disorder. These health problems could be a result of direct poisoning by the polluted land (e.g. eating crops grown on polluted land, drinking water polluted by the leaching of chemicals from the polluted land to the water supply, etc.).

The long term effects of soil pollution are many and can be difficult to deal with, depending on the nature of the contamination.

There are many different ways that soil can become polluted such as:

- Seepage from a landfill.
- Discharge of industrial waste into the soil

- Percolation of contaminated water into the soil
- Rupture of underground storage tanks
- Excess application of pesticides, herbicides or fertilizer
- Solid waste seepage
- The most common chemicals involved in causing soil pollution are:
  - Petroleum metals
  - Pesticides
  - Solvents

Soil pollution happens when these chemicals adhere to the soil, either from being directly spilled onto the soil or through contact with soil that has already been contaminated.

As the world becomes more industrialized, the long term effects of soil pollution are becoming more a problem. Even when soil is not being used for growing food, the matter of its contamination can be a health concern. This is especially so when that soil is found in parks, neighborhoods or other places where people spend time.

Health effects will be different depending on what kind of pollutant is in the soil. It can range from developmental problems, such as in children exposed to lead, to cancer from chromium& some chemicals found in fertilizer, whether those chemicals are still used or have been banned but are still found in the soil.

In the short term, exposure to chemicals in the soil can lead to headache, nausea, fatigue and skin rashes at the site of exposure.

Soil that has been contaminated should no longer be used to grow food, because the chemicals can leach into food and harm people who eat it. If contaminated soil is used to grow food, the land will usually produce lower yields than it would if it were not contaminated. This, in turn, can cause even more harm because a lack of plants on the soil will cause more erosion, spreading the contaminants onto land that might not have been tainted before.

In addition, the pollutants will change the makeup of the soil and the types of microorganisms that will live in it. Certain organisms die because they've lost their food supply. Thus it's possible for soil pollution to change whole ecosystems.

There are some ways to get soil back to its pristine condition or to remove the spoiled soil so that land can be transported to a site where humans won't be exposed to the chemicals, or the soil can be aerated to remove some of the chemicals (which can add to the problem of air pollution, if the chemicals can be released into the air).

Other options include what's known as bioremediation, where microorganisms are used to consume the pollution-causing compounds of chemicals by paring over the tainted area.

None of these are an ideal solution. Preventing contamination of soil is the best way to go. It won't eliminate all potential pollution problems, but choosing to farm organically is a good way to protect the soil from chemicals found in pesticides and other common garden chemicals.

The effects of pesticides on soil micro-organisms can cause a ripple effect that can last for years. Micro-organisms are essential to healthy soil.

Micro-organisms are organisms that are too small to be seen with the human eye. They live on the top-most layer of soil. There are many micro-organisms which live in the soil including:

- Bacteria
- Fungi
- Algae
- Protozoa

Micro-organisms are responsible for the decomposition and recycling of organic materials in the soil. They aid in the plant's absorption of essential nutrients. An example of this is the nitrogen fixing bacteria, *Bradyrhizobium*, which lives in a nodule on the soybean plant. It provides nitrogen to the plant and boosts growth.

Bio pesticides are micro-organisms that can help a plant defend it against pests. These micro-organisms include antimicrobial metabolites, antibiotics and extracellular enzymes.

Pesticides are designed to kill bugs that are harmful to plants. Pesticides kill specific pests on plants such as slugs, beetles and flying insects. The chemicals used in most pesticides can kill the helpful organisms that live in the soil. Some of these chemicals can remain in the soil for years, effectively keeping necessary micro-organisms from working the soil.

Common chemical pesticides that are used in gardens and be large-scale crop producers include the following:

- Basic-Copper Sulfate
- Silica Gel
- Sodium Fluoride
- Carbon Disulfide
- Hydrogen Cyanide
- Methyl chloroform
- Fenthion
- Boric Acid

In places where the chemicals are used extensively, plants will no longer grow at all, or will fail to thrive.

Unfortunately, many pesticides kill more the just intended targets, namely the necessary micro-organisms living in the soil that break down organic material and aid in plant growth. It can take years before micro-organisms can once again live in soil that has had toxic chemicals applied to it.

The use of organic pesticides can keep a healthy balance in the soil. Many organic pesticides are made of minerals or other plant materials that will keep pests at bay and break down quickly in the soil. Examples of some common organic pesticides include the following:

- Cayenne pepper spray–Can be sprayed on the leaves of plants to deter harmful insects.
- Soap spray powder–A spray can be made from the finely ground tobacco leaves and water. It is used to kill sucking insects on plants such as aphids, thrips and spider mites.
- Pyrethrin–Made from the chrysanthemum plant. This organic pesticide is used to knock out flying insects and ground pests such as grubs.
- Neem–Derived from the neem tree. Used to control gypsy moths, leaf miners, mealy bugs, white flies and caterpillars.
- Sabadilla–Derived from the sabadilla lily. Used to control caterpillars, leaf hoppers, stink bugs and squash bugs.

Some the main effects of soil pollution are:

- Disturbance in the balance of flora and fauna inhabiting in the soil.
- Contaminated soil decreases soil fertility and hence there is decrease in the soil yield.
- Reduced soil fertility, hence decrease in soil yield.
- Loss of natural nutrients in soil.
- Reduced nitrogen fixation
- Loss of soil and nutrients
- Increased soil erosion
- Imbalance in the flora and fauna of the soil
- Creation of toxic dust
- Foul odor due to industrial chemicals and gases
- Alteration in soil structure can lead to death of organisms in it
- Reduction in soil fertility
- Loss of the natural nutrients of the soil
- Imbalance in the flora and fauna of the soil
- Crops grown on polluted soil cause health problems on consumption
- Soil pollution creates toxic dust
- Foul odor due to chemicals gases can lead to problems like headaches, nausea, etc.
- Pollutants in soil cause alteration in soil structure, causing death of many soil organisms. This can affect the food chain

Effects on humans

- Soil pollution has major consequences on human health. Consumption of crops and plants grown on polluted soil cause health hazards. This could explain small and terminal illness.
- Long term exposure to polluted soil affects the genetic make-up of the body and may cause genital illnesses & chronic health diseases.
- Chronic exposure to heavy metals, petroleum, solvents and agricultural chemicals can be carcinogenic.
- Exposure to benzene for a long term is associated with higher incidence of leukemia. Mercury causes higher incidences of kidney damage. Cyclodienes are linked to liver toxicity.
- Organophosphates can lead to chain of responses leading to neuromuscular blockage.
- Chlorinated solvents induce damages to liver, kidney, depression of the central nervous system.

Effects on plant growth

- The balance of ecological system is affected due to contamination of the soil.
- Plants are mostly unable to adept to the change in the chemistry of the soil in short time period.

- The microorganisms found in the soil decline and create additional problems of soil erosion.
- Fertility of the soil decrease due to soil pollution, making it unsuitable for agriculture and local vegetation to survive.
- Polluted lands cannot support most forms of life.

Effects On soil fertility

- The chemicals present in the soil due to pollution are toxic and can decrease the fertility of the soil, thereby decreasing the soil yield.
- Agriculture on contaminated soil produces fruits and vegetable that lack quality nutrients.
- Consumption of these may be poisonous and cause serious health problems to people consuming them.

Toxic dust effects

Emissions of toxic gases & foul odor from the landfills pollute the environment and cause odour nuisance.

Contaminants distribution in soils

The distribution of contaminants released to soils by human activities is related to how and where they are added. For industrial-contaminants in the soils of an industrially-contaminated site may vary depending on the activities conducted on the site. The movement of air and water will also affect how soil contaminants move throughout a site. Chemicals may be carried by winds and deposited on the surface of soils; tilling can then mix their surface deposits into the soil. The movement of groundwater of surface water may also affect how contaminants spread from the source.

Many pesticides and soil amendments used for agricultural, industrial, or commercial activities may be found in residential soils. This could happen if former industrial or agricultural lands are later used for residential properties and contaminants remain in the soil spills, runoff, or aerial deposition of chemicals used for agriculture or industry can also result in contamination of the soils of residential sites.

For example, arsenic and lead were once used as pesticides on a number of crops, including orchards. Sodium arsenate was also commonly used on potato crops. Therefore, old orchards, farms, and adjacent areas are places where testing for arsenic and lead might be advisable. Within an orchard, the distribution of these contaminants may be very spotty since individual trees may have been treated, resulting in higher residues under each tree. Collecting multiple soil samples from such an area would help to determine the pattern of contamination.

Due to the wide array of contaminants, soils and site conditions, the levels of possible contaminants will depend on the specific conditions of a particular property. Soil testing can help provide more information about the levels of a particular contaminant (or contaminants).

Processes of Metals Transformation in Soil is shown below in figure.

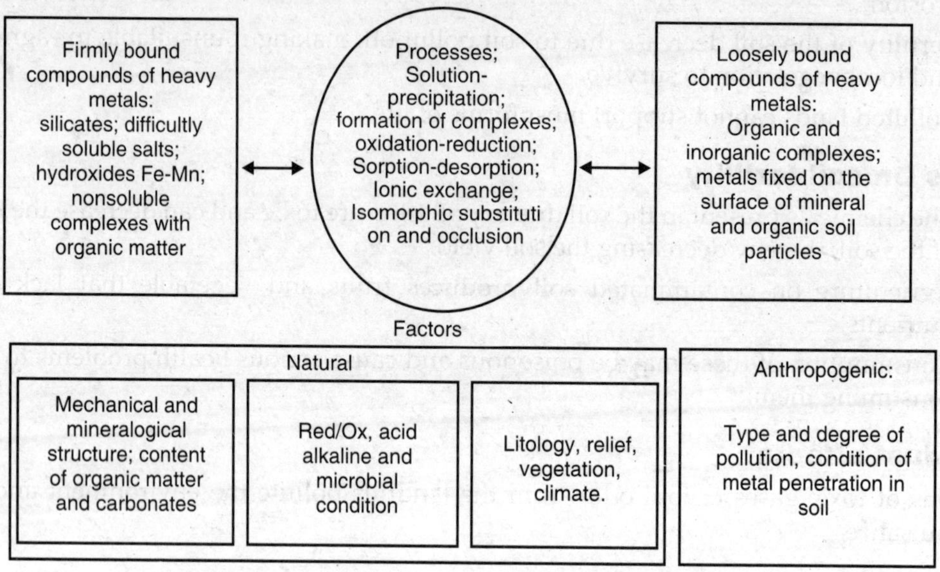

Fig.3.1F_2: Processes of Metals Transformation in Soil

Distribution pattern of metals compounds in soil is shown below in figure.

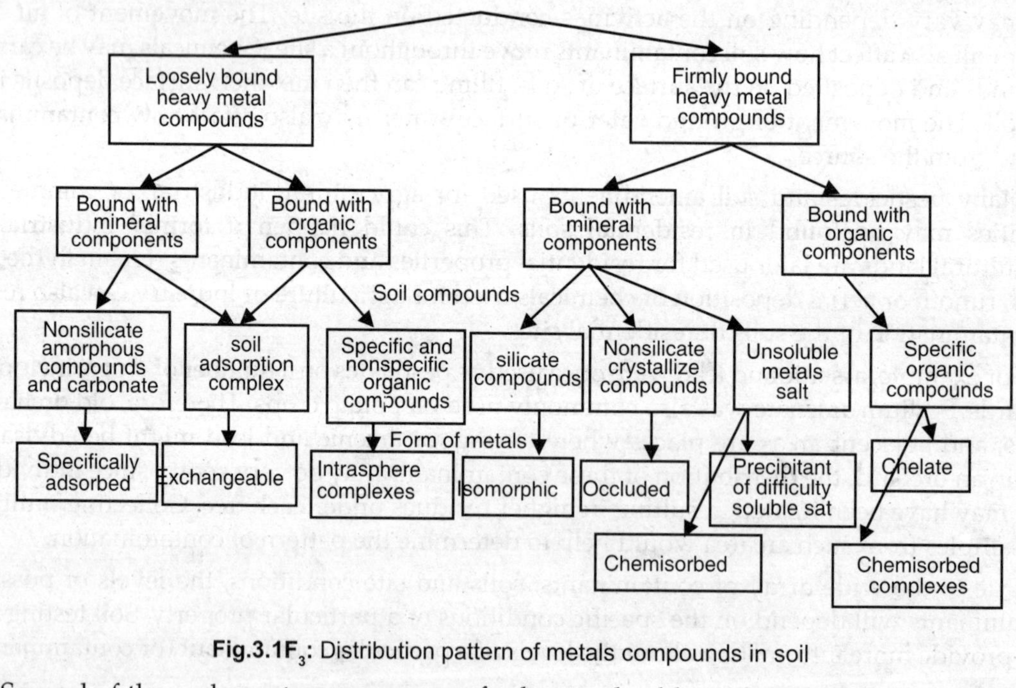

Fig.3.1F_3: Distribution pattern of metals compounds in soil

Several of these elements are necessary for human health, and are beneficial when taken in to the body in foods or as supplements at appropriated, low levels. Conversely, cadmium, lead & mercury have no known biological function and are toxic to humans.

Soil acts as a repository for many heavy metals that human activity releases into environment. This may protect the wider environment to some extent by 'locking away' heavy metals and preventing them reaching others parts of the environment, such as water supplies, However, the soil itself may then present a risk to those who live or eat food grown on it. Some soils have naturally high levels of heavy metals and, in some cases, plant species able to take up and store large amounts of heavy metals evolved in these locations. Human activity such as mining, smelting, industry, agriculture and burning fossil all contribute to the burden of heavy metals in soils, as does our disposal of materials containing heavy metals, a long list which includes municipal waste, paint electronic waste, and sewage.

Our understanding of how heavy metals in soils lead to health risks is limited, compared to our knowledge of impacts via or water.

The bioavailable portion is the amount of a substance that can cause direct effects on plants, animals or humans because it can be taken up by their bodies. Usually, not all of a contaminant found in soil is biologically available. The bioavailability of a contaminant depends on many characteristics of the soil and of the site. Site condition affect how tightly the contaminant is held by soil particles and its solubility (how much of it will dissolve in water). Greater solubility usually means that more of the contaminant is bioavailable, but this also means that the contaminant is more likely to leach out of the soil. Certain chemicals show an ''aging effect' and can become less bioavailable the longer they remain in soils.

Most commonly available soil tests measure a large part of the total amount of a particular contaminant in the sample. Not just the bioavailable portion. The bioavailable portion may be only a small fraction of the total amount. Changes in site conditions, such as soil acidity or organic matter content, can change the bioavailability of a contaminant. There is no easy way to know what portion may be bioavailable. Using bioassay tests to measure uptake of contaminants by plants of soil organisms is the most direct way to estimate bioavailability. Unfortunately, bioassay tests are slow & expensive and are not generally available. For this reason, only the total levels or chemically extractable amounts (commonly used to approximate the total amount) of a particular contaminant are usually measured.

4 CHAPTER Effects of Land/Soil Pollution

FIRST EFFECT OF LAND POLLUTANTS

- Washed away: might accumulate somewhere
- Evaporates. can be a source of air pollution
- Infiltrates through the unsaturated soil to the groundwater
- DDT: fat soluble, stored in fatty tissues
  - Interferes with calcium metabolism
  - Results in thin egg shells in birds
- Agent orange: code name for one of the herbicides and defoliants (results in leaf fall) used by the U.S. military as part of its herbicidal warfare program, During the Vietnam War between 1962 and 1971, the United States military sprayed 80,000,000 litres of chemical herbicides and defoliants in Vietnam.
  - Causes anti fertility, skin problems, cancer.

HEALTH CONSEQUENCES OF LAND POLLUTION

- Land Pollution affects humans and wild life in number of ways such as;
  - It can cause Respiratory problems.
  - Leads to birth defects
  - Causes skin problems
  - Causes cancers
- The toxic substance causing soil/land pollution can get into human body directly by skin contacts.
- Toxic substances being washed into water sources like rivers & reservoirs, pollutes underground water courses.
- Consumption of vegetables and fruits grown on polluted soils.
- Inhalation of Polluted dust or particles.

Cancer

Many of the widely used pesticides on agricultural lands are potentially carcinogenic (Cancerous). The U.S. Environmental Protection Agency (EPA) has identified 15 such chemicals from the 27 most commonly used.

Neurological Problems

The health effects of soil pollution, of course, vary with the contaminant. Some toxins, such as lead, are both naturally occurring and come from man made sources. Children are most vulnerable to the health effects of lead in the soil. Their developing neurological systems put them at risk for serious health issues, causing possible brain and nervous system damage.

Tissue Damage and Irritation

Mercury levels in the soil can increase through deposition from precipitation. Mercury can cause irritation of your internal & external tissues, such as your skin and the lining of your gastrointestinal tract. The severity depends upon the toxin and length of exposure.

Reduction of Soil Health

Soil pollution also degrades soil health, impacting both the environment as well as people in the long term. The same chemicals that can harm people may also affect plants and micro-organisms. Acid rain, for example, can create toxic condition by mobilizing what might otherwise be harmless chemicals in the soil. These toxins in turn can slow growth as well as impact the soil bacteria and other organisms necessary for plant growth.

Detrimental Effects of Land Pollution: Land pollution is the result of urban technological revolution and exploitation of every bit of natural resources. A number of sources are contributing individually or together to land pollution. The effect of land pollutants from different sources on man, animal and vegetation, are discussed below:

(a) Effects of Industrial Pollutants:

- Industrial wastes containing toxic metals (e.g., Hg, Pb, As, Cr, Cu, etc.) kill bacteria and beneficial micro-organisms in the soil. Heavy metals precipitate phosphatic compounds and catalyze their decomposition.
- Industrial wastes containing soluble salts cause crop soil loss, metallic corrosion etc. According to Food Agriculture Organization (FAO), 50% of the irrigated farms in the world are damaged by soluble salts.
- Industrial effluents causing high acidity or high basicity of the soil cause severe agricultural crop damage.
- The accumulation of wastes containing toxic pollutant like oils, greases, plastics, plasticizer, non-biodegradable matter etc. transfers pollutants into food chain causing a number of undesirable changes in living organisms.
- Direct discharge of industrial effluents through sewage system causes poisoning of biological purification mechanism of sewage treatment inducing several soil borne and water borne diseases.
- When discarded plastic materials, textiles, packaging's synthetic fibres, waste paper etc. in soil are burnt, these emit highly toxic gases like HCl fumes, SO_2, NO_x etc.
- Some industrial wastes also contain pathogenic bacteria, which cause significant health hazards. For example, the pathogen Anthrax bacilli are present in tannery waste.

(b) Effects of Urban Wastes:

- The dumped urban wastes along highways and other places in cities spread several chronic diseases posing a serious threat to human health.
- The dumping of waste including building materials sludge, dead animal skeleton and thrown away garbage at public places causes a lot of problems in daily life.
- Solid wastes result in offensive odour and cause clogging of ground water filters.
- Sewage is an excellent medium for the growth of pathogenic (disease causing) bacteria, viruses and protozoa.
- Due to inadequate sewage disposal, refuse mix with solid particles of land. Such type of mixing, damages crop and decreases crop production & soil fertility.

(c) Effects of Agricultural Wastes:

(i) Effects of Fertilizers:

- The presence of excess of NPK fertilizers declines protein of corn, maize, gram & wheat crops and also favours the degradation of carbohydrates & proteins.
- The fertilizers inhibit the synthesis of Vitamin-C and carotene in vegetables & fruits.
- The excessive use of nitrogenous fertilizers in land leads to accumulation of nitrate in the soil which subsequently enters into human through food chain. In human body nitrates are converted into nitrous compounds which are capable of inducing stomach cancer.
- The vegetables & fruits obtained from highly fertilized land are prone to pest, insects and diseases.
- The cereal crops like maize, jawar & pearl millet, grown on alkaline soil absorb higher quantity of fluorides and are responsible for fluorosis.

(ii) Effect of pesticides:

- The pesticides present in the soil sometimes appear in fruits & vegetable and make them unusable.
- Entry of polychlorinated biphenyl (PCB) into human body causes deformities in foetus, nervous disorders, liver and stomach cancer.
- When vegetables contaminated with 0.5 gm or more PCB is consumed, there is possibility of eye damage, darkening of skin and severe ache.
- Pesticides, due to their non-biodegradable nature, are known to seep gradually through the soil into ground water and thereby contaminate the entire public water supply system.
- Herbicides and chlorinated pesticides affect soil texture & function thereby disturbing the natural eco-system.
- Organophosphate pesticides cause muscular weakness, tremors and dizziness in poisoned animals.
- The excessive use has also resulted in defoliation of forests adversely affecting flora and fauna.

(d) Effects of Pathogens in Soil:

- Pathogenic bacteria in soil act as carrier of a number of chronic diseases like cholera, typhoid, dysentery, etc., form soil to man and vice versa.
- Few pathogenic bacteria are Mycobacterium, Leptospira, Pasteurella and Tuberculosis. These cause diseases like amoebic dysentery, cholera, typhoid, polio, hepatitis etc.
- Leptospires (the virus causing leptospirosis) are brought to the soil surface by the excretion of faecal matter or urine of the patients. These have longer life time in soil (few weeks). When a healthy person comes in contact with infected soil, the virus enters into the body through macerated skin or mucous membrane to cause leptospirosis.
- Q Fever is caused by rickettsia coxiella burneti which is present in contaminated dust and soil.
- Antibiosis micro-organisms in soil produce antagonistic effect through metabolic products which may inhibit the growth or cause killing of symbiotic bacteria.
- Certain fungi and actinomycetes present in soil cause most serious subcutaneous and systemic mycoses. The spores of these enter the human body either through respiration or through wounds.
- Algae poison present in soil is considered to be most virulent poison which produces cirrhosis of liver.

(e) Effects of Hazardous Radioactive Pollutants:

- Since radioactive wastes are produced in huge quantities and has high activation energies, these cause a number of hazards to public health system.
- The accumulated radioactive wastes enter into chain by coming into aquatic system and finally enter into living organisms. The accumulation of radioactive substance causes the disruption of physiological process and metabolic change.
- Introduction of radioactive substances into human body through food causes diseases of digestive track.
- It is reported that nuclear explosion at a higher altitude; introduces radioactive wastes into biological cycle and these are distributed in biota present in that region. A burst close to ground makes all the elements of soil potential induced radioactivity.
- The presence of radioactive wastes affects soil quality and soil fertility.

(f) Pollution involves the following mechanism:

- Deposition of Solid waste
- Accumulation of non-biodegradable materials
- Toxification of chemicals into poisons
- Alteration of soil chemical composition (imbalance of chemical equilibrium to soil medium) . By as much, land pollution of this amass globally, it has:
  - Loss of 6 million hectares of land per year
  - Loss of 24 billion tons of topsoil per year
  - Loss of minimum 15 million acres prime agricultural land
- Desertification of land results in the loss of 16 million square miles. The causes for such devastation are generally due to two forms:

Unhealthy soil management methods

- Improper tillage of soil in which excessive tillage result in the deterioration of soil structure.
- Non-maintenance of a proper supply of organic matter in the soil from the imbalance of it especially nitrogen, phosphorus & sulfur unplenished supply after cultivation of vegetation, leaving the soil to stabilize the soil physicality which ultimately lead to desertification.
- Irregular maintenance of a proper nutrient supply of trace elements gives rise to the use of excessive non-biodegradable and accumulate in the soil acidity which eventually destroys useful organisms such as bacteria.
- Improper maintenance of the correct soil acidity which ultimately disrupt the adaptation of various crops and soils as the solubility of minerals present will be affected. In a more acidic soil, minerals tend to be more soluble in rainfall while in alkaline soil, minerals are more insoluble which form complex minerals unable to be at physiological usage.

Improper irrigation practices

- Poorly drained soil result in salt deposits leading to high soil salinity that inhibit plant growth.
- Unirrigated land giving rise to stagnation of agriculture waste products which accumulates and increase land pollution.
- Irregular irrigation leads to decreasing moisturization of land for soil medium and replenishments of solvents.

Soil Pollution Effects on humans

Soil and water Pollution

- Causes cancers including leukemia.
- Lead in soil is especially hazardous for young children causing developmental damage to the brain.
- Mercury can increase the risk of kidney damage; cyclodienes can lead to liver toxicity.
- Causes neuromuscular blockage as well as depression of the central nervous system.
- Contact with contaminated soil may be direct (from using parks, schools etc.) or indirect (by inhaling soil contaminant).
- Soil pollution may also result from secondary contamination of water supplies and from deposition of air contaminant.
- Since it is closely linked to water pollution, many effects of soil contamination are similar.

Effects of soil Pollution on Animals

- Can alter metabolism of microorganisms and arthropods in a given soil environment; this may destroy some layers thus have a negative effect on predator animal species.
- Small life form may consume harmful chemicals which may then be passed up the food chain leading to larger animal mortality rates and even animal extinction.

Soil Pollution Effects on Trees and Plants

- May alter plant metabolism and reduce crop yields.
- Trees & plants may absorb soil contaminants and pass them up the food chain.

Salination of Soil

Increase in salt concentration of soil reduces its productivity and its quality. Inadequate drainage especially in irrigated areas leads to accumulation of salt at the soil surface. During summer season salts from lower layer move to soil surface. In India intensive farming with poor drainage has increased soil salinity.

Some specific soil Pollutants and their Effects

- **Fluorides**

  They combine with Mg_2+ of chlorophyll and hence inhibit photosynthesis, cause abscission of fruit. In human beings mottling of teeth (fluorosis) is an indication of fluorination. Bone fluorosis results in weak bone of knee.
- **Nitrogen fertilization (nitrates + nitrite)**

  Toxic concentration in leaves and fruits enters food chain. In alimentary canal, its activity changes nitrates into nitrites & combine with hemoglobin and so oxygen transport is reduced. It causes cyanosis (blue babies due to bluish tint of skin) . Nitrate poisoning is fatal unless methylene blue is injected which produces breathlessness.
- **Weedicides**

  They are usually metabolic inhibitors which stop photosynthesis and other metabolic processes thereby killing the plant. Some cause phloem cells to block transport of organic food.

Soil Pollution control

- Decrease the level of volatile constituents in petroleum products that are absorbed to soil and dissolved into groundwater; injection of contaminant free-air into subsurface saturated zone, enabling a transition state to a vapor.
- In combination with soil vapor extraction (SVE) which creates a negative pressure through a series of extraction with migration.
- Effective in decreasing levels of volatile organic compounds (VOCs) which can either use vertical or horizontal sparge.
  - Vapor/dissolved phase partitioning of the constituents the equilibrium distribution of a control phase and the vapor phase.
  - Permeability of soil determines rate of air injection into the saturated zone.

Low Temperature Thermal Desorption

- Ex situ technology which uses heat to split petroleum hydrocarbon from excavated soils through physical methods.
- Effective in decreasing levels of petroleum products like gasoline, jet fuels kerosene, diesel fuel, heating oils and table.
- Mechanism criterion: contaminated soils excavated and transferred to stationary facilities; mobile units operate.

- Treatment modes:
  - Rotary dryers
  - Asphalt plant aggregate dryer
  - Thermal screw
  - Conveyor furnace

Soil Vapor Extraction (SVE)

- In situ technology in reducing the level of volatile constituents in petroleum products absorbed to soil in the vacuum are introduced to wells near the source of pollution, where the evaporated vapors are drawn to the extraction/ absorption before being released to atmosphere.
- Effective in use of Volatile Organic Compounds (VOCs) & certain semi volatile organic compounds (SVOCs).
- Efficiency of SVE is determined by:
  - Permeability of soil
  - Soil structure and stratification
  - Soil moisture
  - Depth to groundwater

Soil Washing

- Elimination of Hazardous waste and decrease the concentration to a minimum, which spilt the contaminated fine soil (sand and gravel).
- Upon completion, majority of fine silt and clay particles are further treated by bioremediation or disposed.
- Effective with low presence of silt & clay, metals, pesticide and fuel soils, with combinations of other treatment.
- Important treatment mode is the soil scrubbing unit.

Agriculture measures to Control soil pollution

- Use of pesticides should be minimized.
- Use of fertilizers should be judicious.
- Cropping techniques should be adopted to prevent growth of weeds.
- Special pits should be selected for dumping wastes.
- Cropping techniques should be improved and forest management.
- Wind breaks and wind shield in areas exposed to wind erosion.
- Planning of soil binding grasses along banks and slopes prone to rapid erosion.
- Afforestation and reforestation.

APPENDIX 4.1

Effects of soil pollutants on plants

Soil Pollution and Plants

Most of the atmospheric pollutants finally come down and settle on the ground surface, causing soil pollution. Pollutants also come back to soil when polluted water is used for irrigation. Many pollutants are directly deposited on the soil as sewage, industrial effluents and chemical fertilizers. The soil pollution by various pollutants has far reaching effects on plants and vegetation.

(a) Effects of soil acidification on plants

The oxides of sulphur & nitrogen, chlorides, fluorides, ammonium etc. emitted into the atmosphere in combustion from various industries come down as dry or wet deposition (acid rain) onto the soil and lower the soil pH. Increased acidity of soil results in following effects:

- The activity of soil microbes, particularly of decomposers, is reduced. The decomposition of organic matter and consequently nutrient cycling in the soil is reduced. It ultimately adversely affects the growth of plants.
- The bases in the soil are leached down due to soil acidity. As exchangeable bases become deficient in the soil, plant growth is reduced due to nutrient deficiency.
- The roots, particularly the root hairs, are damaged resulting in reduction in nutrient uptake by plants. Increased acidity mobilizes heavy metals like Al, Cd, Zn, Hg, Mn, Fe, etc. These spread rapidly in the soil along with soil water and reach concentrations toxic to plants. Consequently, plants show species specific metal toxicity symptoms. Al toxicity generally damages root hairs and reduces nutrient uptake while Fe toxicity has general adverse effect on plant growth.
- In some soils, acidification increases weathering of silicate minerals destroying the mineral structure of the soil. This leads to poor growth of vegetation in general.
- In some marginal soils and grasslands, acidification increases the supply of plant nutrients like sulphur & nitrogen. The vegetation is thus, benefited by soil acidification and plants may show better growth.

(b) Effects of pesticides on plants

Various pesticides, insecticides, fungicides etc. are used in agriculture as foliar spray of area applied to soil far in excess to the requirement causing soil pollution. These substances pollute the soil depending upon their volatility, biodegradability persistence, leaching, reactivity and adsorption on the soil particles. Many of these exchangeable charges, are adsorbed on silicate clay micelle of humus molecules on the pH-dependent exchangeable charges sites and are later absorbed by the plants.

Absorbed pesticides substances produce characteristic-species-specific toxicity symptoms in plants just like their aerial overdose.

Fungicides reduce abundance of soil fungi and actinomycetes and interfere with decomposition of soil organic matter adversely affecting the nutrient cycling.

Pesticides increase the abundance of predator soil microbes and consequently increase the abundance of their prey species.

In general, species composition of the soil microflora and fauna is changed by pesticide substances.

The inorganic pesticides contain arsenic and sulfur. These substances cause symptoms like yellowing, necrosis, shot holes, premature defoliation etc. in plants.

(c) Effects of herbicides on plants

Various herbicides are used for weed control in agricultural practice. The symptoms in response to a particular herbicide may be characteristic but their development depends upon the dose to which plants are exposed, rate of growth of plant, weather conditions and the plant species. General effects of some common herbicide substances on the plants are as follows:

- Acetanitides (e.g. Prochlor, Metachlor) cause stunting of plants exposed, rate of growth of plant, weather conditions and the plant species.
- Acetanilides (e.g. Prochlor, Metachlor) cause stunting of plants and roots. In brassicas, these cause yellow, red and blue colouration of leaves with pronounced stunting.
- Amides and carbamates (e.g. Diphenamide, Propzyamide, Chlorogopham and Asulam) cause stunting without chlorosis of leaf-scorch in mild dose. Crops generally become greener in colour. Roots show thickening and stunting. In the cereals, coleptile becomes stunted and swollen.
- Benzo nitriles (e.g. Dichlorobenzil) cause stunting of plants and thickening of roots. In shrubs, bark at the ground level develops necrosis. In some cases, marginal yellowing occurs in the leaves.
- Aliphatic acids (e.g. TCA, Dolpon) cause stunted growth of shoot, loss of leaf wax and mild irregular chlorosis. Growth regulators (e.g. 2, 4, D, MCPA, Chlopyralid, and Mecoprop) cause hormone-type distortion in leaves and fruits. In the brassicas, stem splitting occurs. In the cereals, ears become distorted, becoming blind or with shriveled grains.
- Ureas and Uracils (e.g. Diuron, Linuron, Bromocil, Lencil) cause inhibition of photosynthesis leading to yellowing along the veins that later extends across whole leaf. In the cereals, chlorosis starts in the middle of leaf but extends quickly to the tip.
- Sulphonyl ureas (e.g. Chlorsulfuron) cause inhibition of growth in susceptible plants and these ultimately die due to root growth being stopped.
- Triazinones (e.g. Metribuzin) cause inhibition of photosynthesis that leads to yellowing and tip-scorch in the leaves.
- Triazines (e.g. Atrazine, Simazine) cause chlorosis & necrosis spreading toward tips of leaves and inhibit photosynthesis. This is more severe in older leaves.
- Thioarbamates and ethofumesates cause chlorosis, stunting & leaf-curling. Younger leaves become stuck to older leaves. In the cereals, this condition often leads to characteristic pink colouration at the base of the stem.

(d) Effects of sewage and ash pollution on plants

Sewage matter is commonly used as fertilizer or deposited as waste on the soil. Effects of such pollution are mostly common to all plants. The organic matter in sewage decomposes & produces nitrogenous substances that become excessive in the soil and harm the vegetation. Decomposition of sewage may also release various toxic heavy metals that cause characteristic heavy metal toxicity symptoms in plants. Detergent substances may also be released from sewage causing characteristic injury to plants.

Ash produced mainly from combustion of coal in thermal and industrial plants used for landfilling or deposited on soil makes the soil unfit for vegetation. It may release many toxic substances in the soil causing characteristic plant injuries.

(e) Effects of chemical fertilizers

Chemical fertilizers are generally used far more than the requirements of the crop. The unutilized fertilizers cause soil pollution.

Toxic concentrations of nitrogen fertilizers cause characteristic symptoms of nitrite or nitrate toxicity in plants, particularly in the leaves. Nitrogenous fertilizers generally cause deficiency of potassium, increased carbohydrate storage & reduced proteins, alteration in amino acid balance and consequently change in the quality change of proteins.

Ammonium and nitrate produce acids in the soil and increase soil acidity. Nitrate and nitrite bacteria are reduced while ammonifying bacteria are increased in the soil disturbing the nitrogen cycle.

Excessive potash in the soil decreases ascorbic acid and carotene in the plants.

Superphosphates cause deficiency of Cu and Zn in plants by interfering with their uptake.

Excessive lime prevents the release of Co. Ni, Mn & Zn from the soil and their uptake by plants is reduced causing their deficiency symptoms.

Excessive deposition of various substances released from chemical fertilizers into the soil generally causes their over-absorption by plants. These over-absorbed substances become accumulated in plant parts (bioaccumulation).

(f) Effects of industrial effluents on plants

Various inorganic and organic substances are present in the industrial effluents. These substances mostly remain tied up in the soil and are not readily available to plants. However, they affect various soil characteristics.

Effluents affect the mineral structure, soil pH, exchangeable base status etc. of the soil and thus indirectly affect the plants. The pH the soil is disturbed making soil either acidic or alkaline. Various inorganic and organic chemicals are accumulated in the soil up to levels toxic to plants.

In highly polluted soils, plants absorb and accumulate toxic substances (bioaccumulation). These substances may or not produce direct injury symptoms in plants but are passed on the higher trophic levels (bio magnifications).

(g) Effects of radioactive pollutants on plants

A variety of radioactive waste materials like Strontium-90, Cesium-137, lodine-131, Plutonium, Uranium, Americium, Curium, Neptunium etc. are added to soil nuclear activities. These substances may be washed into water or may be directly added to water that is used as coolant in nuclear power plants.

Uptake of radioactive substances by plants depends upon pH and organic matter status of the soil. Various radioactive materials show different solubility and absorption by plants. Plant uptake is generally highest for Strontium-90 which behaves as Ca in the soil and is absorbed by plants like it. Cesium-137 behaves like K but its uptake by plants is very limited.

Absorbed radioactive substances are generally in the vegetative parts of the plants, particularly the leaves. Their accumulation in crop fruits seeds is very low. Accumulation of absorbed radioactive substances in the plants may be up to 100 times their concentration in the soil of water.

Radiation from radioactive substances may also adversely affect the plants. Pines are eliminated on exposure to 20-30 roentgen / day while most plants die at 200 roentgens/ day. Only lichens and mosses are highly resistant to radiation.

Radiation also damages chromosomes. It increases the frequency of chromosomal aberrations and causes genetic mutations. Such genetic changes may adversely affect plant metabolism or change their characteristics in subsequent generations.

Interrelationship among acid Deposition, Plants, and Soil is shown below in figure.

- Nutrient leaching
- Heavy metal release
- Weakens trees

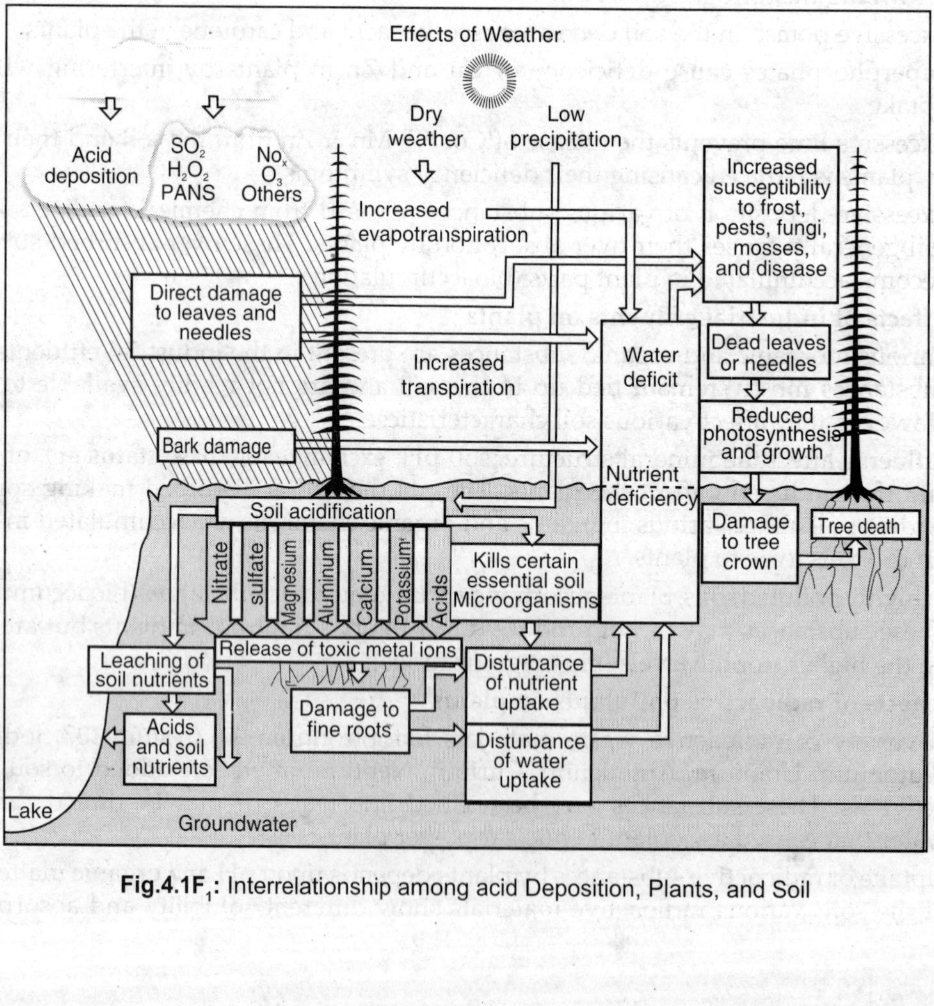

Fig.4.1F$_1$: Interrelationship among acid Deposition, Plants, and Soil

Soil Management

There are around 16 micro nutrients needed for the healthy growth of a plant. During each harvest, the soil's fertility will keep progressively reducing. To protect the fertility of the soil, it has to be tested each time after harvesting. A careless application of fertilizers will only further deplete the soil fertility. It is also advisable to apply micro and macro nutrients simultaneously into the soil, cropping would help to balance the nutrients in the soil.

Although the problems of soil management vary according to the soils & their situations and the climatic conditions, there are fundamental factors which govern the choice of soil management practices. The productive capacity of the soil should be improved and maintained by providing adequate organic manure & plant nutrients.

Soil Test

For maximum production and rational soil, a knowledge of fertility status & the physical properties is essential. Soil testing is one of the methods of determining the fertility status of the soils, so that recommendation in regard to deficient nutrients or soil amendments can be made.

To collect soil samples

Remove all leaf litter and debris from the top soil and using a soil tube, an auger, a spade or a pick-axe, dig the soil in a V shape up to a depth of 10 cm and collect the sample. Dry the sample. The soil be removed from the centre of the land but manure filled or compost areas have to be avoided.

Given Below are 12 micro and macro nutrients essential for plant growth.

Table 4.1T_1: Functions of various plant nutrients

S.N.	Plant Nutrients	Functions
1	Nitrogen	For leaves, stem growth, seed development
2	Phosphorous	For roots to spread and absorb water
3	Potassium	For protection from disease, pest resistance, and compensating evapotranspiration
4	Iron	For preparing chlorophyll and for energy development & transfer
5	Zinc	Growth stimulator
6	Manganese	For absorption of CO_2 and for transfer of nitrogen
7	Copper	For preparation of primary protein
8	Boron	In cell formation and division
9	Molybdenum	To absorb nitrogen and to resist excess micro-nutrients
10	Sulphur	For chlorophyll and root development
11	Calcium	In cell development and for transfer of food
12	Magnesium	For protein preparation-as an enzyme catalyst

Other nutrients such as carbon, hydrogen, oxygen and chlorine too play vital roles in plant growth.

Land is very valuable but limited resource, as the population increases rapidly, many highly urbanized cities are faced with acute space problems, as in Calcutta or Bombay. Besides the limited availability of land, 175 million hectares of land are becoming less productive every year. India loses 20 tons of topsoil per hectare in a year due to floods, rainfall and deforestation. 30 to 50 % of lands under irrigation can go out of cultivation at this rate because of water logging and salinity.

The scenario of desertification is compounded by pollution which includes:

- Indiscriminate discharge of industrial effluents on land and into water bodies.
- An increase in the use of fertilizers for agriculture.
- Open defecation by animals and human beings.
- Accumulation of solid waste; this is a major problem in developed countries like India where the garbage and refuse products are not degraded.
- Radioactive substances from nuclear explosions which are released into the soil.

Major Metallic Soil Pollutants and Their effect on Human Health are tabulated below

Table 4.1T$_2$: Metallic Soil Pollutants and Their Effects on Human Health

S.N.	METAL	SOURCES	EFFECTS
1	Arsenic	Occurs Naturally	Chronic poisoning leads to loss of appetite and weight, diarrhea, alternating with constipation, gastro intestinal disturbance, with constipation, gastro intestinal sometimes skin cancer.
2	Cadmium	Mining, metallurgy chemical industry and electroplating	Lead to chronic poisoning and affects the proximal tubules of the kidney, causing formation of kidney stones.
3	Lead	Lead smelters storage battery	Lead poisoning can lead to severe mental retardation or death.
4	Mercury	Industrial wastes	Methylmercury compounds are much more toxic than other forms of mercury, causes neurological problems and damages renal glomeruli and tubules.
5	Cyanides	Waste from treatment of metals, dismantling of electroplating shops, etc.	Rapid death may follow due to exposure to cyanide as a result of inhibition of cellular respiration.

APPENDIX 4.2

Effects of specific soil pollutants on human health

Soils vary considerably in their composition across the world regions which is significant to human health, because parent material (the weathered rock materials from which soils are formed) topography, climate, organisms and time will lead to soils with different physical & chemical properties. A soil's unique composition will tell how much water it can hold, the living organisms it supports, which chemical reactions are likely to occur, and how it cycles nutrients. All of these factors will determine what happens to potentially harmful contaminants in soils, how they may be transported or transformed, and the extent to which they may be available in chemical forms that are harmful to human health.

Soil pH (acidity) is of particular importance because it controls the behavior of metals and many other soil process. Heavy metal cations (positively charged metal atoms) are most mobile in acid soils. This means that metal contaminants are more available for uptake by plants, or to move into the water supply. Making soil less acidic, by adding lime, is one way to reduce the bioavailability of metals.

A healthy soil has several physical, chemical and biological properties: it needs to incorporate adequate organic matter, have a good structure, and be home to a diverse mix of organisms. These properties allow the soil to carry out important functions, and may be achieved in a natural setting by a soil to improve the soil's health. Agricultural soil health is linked to human health; as poor soils yield fewer crops with decreased nutritional value. Healthy soils also limit erosion, and help improve air & water quality.

Contamination can seriously affect soil's ability to perform some of its key functions in the ecosystem. Soil is a living resource, but once contamination exceeds a certain threshold, the soil may be considered functionally dead. Pollution by heavy metals & many organic contaminants is practically irreversible.

Routes from soil to human intake: Soil can enter into our body by three main routes ie, eating (ingestion) inhalation and through skin.

Eating (ingestion): Eating soil is a widespread practice. Small children are very likely to eat soil while playing outdoors. Young children are at highest risk from contaminated soils (for example, children absorb lead via their digestive system five times more efficiently than adults). Accidental ingestion may occur in adults (for example, by eating vegetables with some soil still attached), but in some parts of the world, adults also deliberately eat soil for a number of cultural reasons. Direct ingestion is the most important pathway for human exposure to soil contamination, although other pathways have same importance in certain situations.

When consumed, some chemicals are absorbed through the lining of the mouth, while others are swallowed and move into the digestive system. From here, they may be absorbed into the body and transported to the liver. Once in the liver, some chemicals are largely returned to the digestive system via bile, but others will enter the bloodstream. Some chemicals are broken down to a certain extent in the liver before they reach the blood. Where chemicals are not absorbed, and remain in the gut, they generally do not cause an adverse response, unless they have some direct toxicity to the gut lining.

Inhalation: Working with soil (for example, in agriculture) releases into the air chemicals that may be inhaled by workers and others nearby. Very small particles may lodge in the lungs, and there is a chance that contaminants may be absorbed into the blood stream. Compared to ingestion, this is a far less significant source of exposure, but may be relevant to those exposed repeatedly over a long time period.

Skin contact: Absorption through the skin tends to favour more organic compounds. This is less of a problem for heavy metals, although some specific forms [Cr(IV) - the more toxic form of chromium, or inorganic mercury] can cause skin contact problems. Absorption of a chemical through the skin is known is 'dermal absorption', or sometimes 'cutaneous' or 'transcutaneous absorption'.

Indirect contact: Soil contaminants may move from soils into ground or surface water, leading to contaminated drinking water. They may also be taken up by plants which are subsequently consumed, either by humans or by agricultural livestock, causing contaminants to enter the human food chain, some of these effects may be quite significant, as in the case of dioxins accumulating up the food chain, or large quantities of cadmium in crops grown in contaminated soil. High levels of arsenic in drinking water supplies are often another significant indirect result of soil contamination. Arsenic may also be naturally present in groundwater.

A contaminant becomes toxic in the human body once the body's own detoxification systems become overloaded. At this point, the body starts to be exposed to excess amounts either of the chemical itself or of a metabolite produced when the body's normal metabolic pathways (the means of processing the toxic compound) are saturated. If a chemical accumulates in tissues, reaching toxicity may be an event that results from long-term accumulation, Factors that are relevant in this case are the body's rate of elimination (by metabolism or excretion), and the overall 'body burden'-the quantity of chemicals stored in body tissues. Reliable data from human populations exposed to known levels of chemicals are not common, with the exception of human pharmaceuticals. For chemical contaminants, levels likely to pose risks to human health are estimated from toxicology studies on laboratory animals, and models.

Types of soil contaminants

The types of contaminant that could be present in soils and affect human health, it is worthwhile to consider the chemicals that offer the greatest threat to human health first. The figure below shows the Chemicals relating to soil which is of major public health concern.

Some of the known health effects from these chemicals are based on case sources other than soil was the cause (e.g. drinking polluted water).

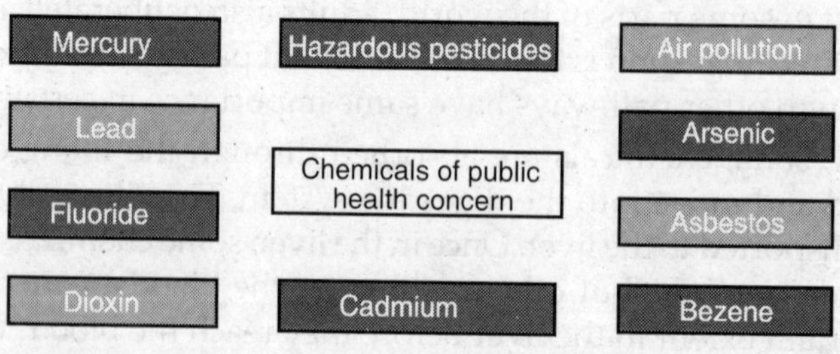

Fig.4.2F$_1$: Ten chemicals in soil of major public health concern

Human health effects of ten major chemicals of concern as per WHO (World Health Organization) as shown above in figure is tabulated below.

Table 4.2T$_1$: Human health effects of ten major chemicals of concern as per WHO

S.N	Chemicals of concern, Sources/uses	In soil?	Used by humans as a nutrient?	Toxic to humans how?	Health effects
1	Air pollution	No	-	-	-
2	**Arsenic** Pesticides; gold, lead copper, nickel, iron and steel mining and/or processing; coal burning; wood preservatives Pharmaceutical and glass industries, sheep dip, leather preservatives, pigment poison bait, agrochemicals, antifouling paint, electronics industries.	Yes	No	Main exposure through consumption of groundwater containing naturally high levels of inorganic arsenic, food prepared with this water, or food crops irrigated with water high in arsenic.	Intake of inorganic arsenic over a long period can lead to chronic arsenic poisoning, (arsenicosis). Gastrointestinal tract, skin, heart, liver and neurological damage. Diabetes. Bone marrow and blood disease. Cardiovascular disease. Carcinogenic. Organic arsenic compounds are less harmful to health, and are rapidly eliminated by the body.
3	**Asbestos** Mining and milling of raw asbestos (historical) for construction and product manufacture. Historical: releases into the air and soil around	Yes	No	Exposure occurs when asbestos-containing material is crumbling or disturbed, releasing microscope asbestos fiber	Some inhaled asbestos fibers reach the lungs, where they become lodged in lung tissue and may remain for many years. This causes: • Parenchymal asbestosis • Asbestos-related pleural abnormalities

	refineries, power plants, factories handling asbestos, shipyards, steel mills, vermiculite mines, and building demolitions, Current repair, renovation, removal, or maintenance of asbestos. Gardening.			into the air and dust. The main route of entry is inhalation, but it can also be ingested or lodged in the skin.	• Lung carcinoma • Pleural mesothelioma Health effects may not emerge for decades, but lung cancer and pleural mesothelioma have high mortality rates. Historical, occupational exposure from manufacturing and construction work is a common cause.
4	**Benzene**	No Benzene is not persistent in surface water or soil, either volatilizing back to air or being degraded by bacteria (unless present in very high quantities).	-	-	-
5	**Cadmium** Zinc smelting, mine tailings burning cola or garbage containing cadmium, rechargeable batteries (nickel-cadmium batteries account for over four-fifths of cadmium consumption), pigments, TVs,	YES Cadmium in soil may enter plant crops (depending on soil characteristics, pH etc.)	No	Cadmium in soil or water used for irrigation can lead to accumulation in plants that enter the human food chain. Cadmium may also accumulate in animals at levels that do not affect the animal's	Liver and kidney damage, low bone density. These symptoms are known as itai-itai disease. First identified when cadmium from mining in the Toyoma Prefecture of Japan led to high levels of cadmium in rice, which accumulated in local people.

	solar cells, steel, phosphate fertilizer, metal, plating, water pipes, sewage sludge.			health, but can affect humans consuming animal products.	Diets poor in iron and zinc vastly increase the negative health effects of cadmium. Carcinogenic (by inhalation).
6	**Dioxin** Including Polychlorinated dibenzofurans (PCDF). Waste incineration reprocessing metal industry, paper and pulp industry, Contaminated herbicides (a major source). Stored PCB-based industrial waste oils (often with large amounts of PCDFs).	YES These chemicals are most commonly found in soil, sediments and food, with low levels in air and water.	No	Human exposure to dioxin-like substance occurs mainly through consumption of contaminated food. More than 90% of human exposure is through food, mainly metal and dairy products, fish and shellfish.	Dioxins are highly toxic and can cause reproductive & development problems, damage the immune system, interfere with hormones and also cause cancer.
7	**Fluoride**	Yes-but is generally immobile.	Yes A micronu-trient: Appropriate levels strengthen teeth.	Usually associated with high levels of fluoride in drinking water.	Skeletal fluorosis fluoride accumulates progressively in the bone over many years. Early symptoms include stiffness and pain in the joints. Crippling skeletal fluorosis is associated with osteosclerosis, calcification of tendons & ligaments, and bone deformities.

8	**Lead** Batteries, solder, ammunition, pigments, paint, ceramic glaze, hair colour, fishing equipment, leaded gasoline (vehicle exhaust), mining plumbing, coal burning, water pipes.	Yes	No	Leaded fuel and mining activities are common causes for elevated lead levels in topsoil.	• Neurological damage • Lower IQ and attention • Hand-eye-co-ordination impaired • Encephalopathy • Bone deterioration • Hypertension • Kidney disease
9.	**Mercury** Electrical switches, fluorescent light bulbs, lamps, batteries, thermometers, dental fillings, mining (particularly artisanal/ small scale gold mining), pesticides, medical waste, burning coal and fuel oil, chlor-alkali industry.	yes	No	Main exposure route for the population at large is via eating contaminated seafood. For children is direct ingestion of soil	• Kidney disease • Central nervous system (CNS) and gastric system damage. • Affects Coordination, eye sight and sense of touch • Liver, heart and kidney damage. • Teratogenic
10.	**Hazardous pesticides** Herbicides derived from trinitrotoluene may have the impurity dioxin, which is highly toxic. Synthetic insecticides, such as DDT (now banned) can still be found in the environment worldwide	Yes	No	Organic pesticides accumulate in the food chain	Organic chemicals, including pesticides, have been linked to a wide range of health problems, but we tend to be exposed to a cocktail of these chemicals at low levels. Conclusive proof of cause and effect in humans is challenging.

Human health effects of Heavy metals present in soil

'Heavy metals' is a term used for elements with metallic properties having specific gravity more than 5. However, it is not an accurate description, since the definition of 'heavy' is not fixed, and some so-called heavy metals, such as arsenic and antimony, are semi-metals or metalloids.

Another description often used interchangeably with heavy metals is 'trace elements'. These elements occur naturally in rocks and in variable amounts in soils, depending on their location and the rocks that have broken down to make the soil's components. The group 'heavy metals' for the purpose of discussing health risks or impacts generally includes:

- Arsenic (As)
- Lead (Pb)
- Cadmium (Cd)
- Chromium (Cr) [although only the form Cr (VI) is toxic]
- Copper (Cu)
- Mercury (Hg)
- Nickel (Ni)
- Zinc (Zn)

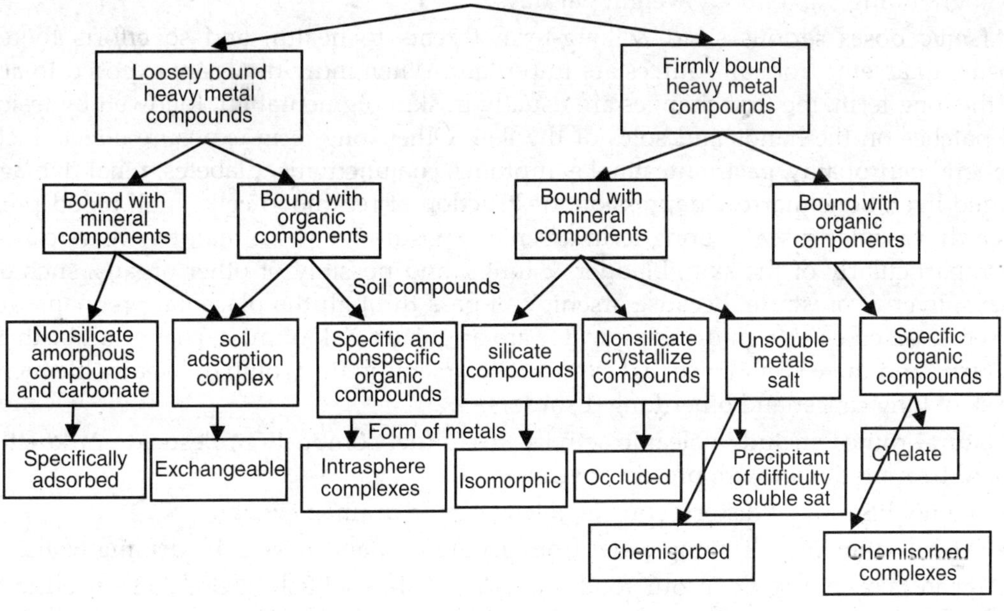

Fig.4.2F$_2$: Distribution pattern of heavy metals compounds in soil

Health impacts of some important soil pollutants is discussed below.

Arsenic

Arsenic is found throughout the earth's crust, generally in the form of arsenic sulfide, or metal arsenates and arsenides. Key industrial applications of arsenic include antifungal wood preservatives (e.g. for railway sleepers), pharmaceutical & glass industries, manufacture of alloys, sheep dips, leather preservatives, pigments, antifouling paints & poison baits, and agrochemical production (particularly for orchards and vineyards). Arsenic compounds are used in small amounts in the optical and microelectronics industries.

Health effects of arsenic exposure

Much of the evidence for the long-term effects of arsenic on human health comes from southeast Asia where there is a natural belt of arsenic alluvium/sediments which were deposited millions of years ago in the Brahmaputra and Ganges river basins. Bangladesh, parts of India, Myanmar & Nepal are all affected, and mining in areas of Thailand has also caused arsenic contamination. An estimated 30 million people may be at risk from arsenic related disease as a result of contaminated water in the region.

Humans are exposed to inorganic arsenic through soil, air, water and food. This includes children ingesting soil, ingesting certain traditional medicines & foods, and ingesting water. In soils in this region, arsenic is present at levels between 0.2 and 40 micrograms per gram (μ g/g) of soil.

The levels of arsenic in food in affected countries vary, but a far greater threat is considered to be arsenic in drinking water. Arsenicosis (sometimes also called arsenism) is caused by prolonged exposure to low, non-lethal doses of arsenic, in the range of 0.005 to 0.09 milligrams per kilogram (mg/kg) of body weight per day.

Arsenic poses serious short & long-term threats to health, and so efforts to reduce exposure to arsenic from all sources are important. When individuals are exposed to arsenic over the long term, the first changes are usually in skin pigmentation, followed by lesions & hard patches on the hands and soles of the feet. Other long-term exposure effects includes peripheral neuropathy, gastrointestinal symptoms, conjunctivitis, diabetes, renal damage, an enlarged liver, bone marrow depression, destruction of red blood cells, high blood pressure and cardiovascular disease. Long-term arsenic exposure for more than ten years can cause cancer, particularly of the skin, bladder & lungs, and possibly of other organs, such as the kidneys, liver & prostrate. Because arsenic can pass through the placenta, pregnant women exposed to arsenic through drinking water are at greater risk of miscarriage, stillbirth & pre term birth, and there is evidence that exposure to arsenic in the womb or in early life increases the risk of lung cancer and other lung disorders.

Natural routes, include volcanic activity, fossil fuel combustion, pesticide production & use, and treating timber with preservatives.

- Remobilisation of historic sources, such as mine drainage water.
- Mobilisation into drinking water from geological deposits, e.g. by drilling wells.

We can take in arsenic in our food, including fish, shellfish, meat, dairy products and cereals. The type of arsenic found in fish and shellfish is usually organic, which has low toxicity. The production of antifungal wood preservatives is a significant industrial source of arsenic that can contaminate soil.

The form that arsenic takes in soils depends on a number of factors, including the soil's pH, and biological activity, where iron, clay & organic matter are present in soils, arsenic's availability becomes restricted. Even where land is contaminated, plants rarely contain much arsenic; cereals and vegetables, especially where soil is sandy, have the greatest concentrations of arsenic.

Natural processes are responsible for polluting wells in locations such as Bangladesh and Taiwan with arsenic, but in other countries, the pollution has a human source. Cornwall in the UK was once the world's largest arsenic producer, and soil in some parts of Cornwall has some of the world's highest arsenic concentrations.

Some human populations have a greater tolerance for arsenic, due to genetic factors that mean they tend to process and excrete arsenic faster than other groups. Initial studies show this to be the case for indigenous people in the South American Andes, who have lived with arsenic contaminated drinking water for thousands of years (as evidenced by samples taken from mummies), and who carry a gene that helps to metabolize arsenic.

Cadmium

Cadmium is a non-essential & toxic element for humans, and has no use for plants or animals either. It damages the kidneys, causing excess production of proteins in the urine. The duration and level of exposure to cadmiums dictates the severity of the effect.

Skeletal damage is another critical effects of chronic cadmium exposure at levels somewhat higher than those where protein in the urine would be an early indicator. Cadmium is also carcinogenic, if inhaled. Mainly stored in the liver & kidneys, excretion of cadmium is slow and it can remain in the human body for decades. Levels of the element tend to build up in most body tissues with age.

Cadmium is associated with skeletal damage, evidenced by low bone mineralization, a high rate of fractures, increased osteoporosis and intense bone pain. These were features of itai-itai disease, first described in Japan in the 1940s among people who had eaten rice grown on fields irrigated with cadmium rich water. A low calcium diet plus high cadmium exposure led to kidney disease followed by bone disease.

Around 90% of cadmium exposure in non-smokers is through food. Crops take in cadmium from soils and the rate of uptake is influenced by factors such as soil pH, salinity, humus content, crop species & varieties and the presence of elements (e.g. Zinc). Some population groups are especially vulnerable to increased exposure and uptake of cadmium:

Vegetarians or individuals who consume large amounts of cereals and pulses are likely to have higher exposure than the general population, as agricultural crops (especially irrigated rice) account for most of the cadmium intake.

Those with a high intake of shellfish and organ meat from animals may have a particularly high intake of cadmium.

People with other nutritional deficiencies may also be at risk.

Smokers: tobacco plants absorb cadmium from soil, as other plants do, and are an important source of cadmium uptake. Non-smokers may also be affected through passive exposure to secondary smoke.

People living in the vicinity of industrial sources and other point sources of cadmium release can be exposed to an increased level of cadmium.

The average weekly intake of cadmium from food in most countries is within the range of 0.7–208 $\mu g/kg$ body weight. Given their smaller size, children may be taking in more cadmium per kilogram of body weight than adults.

In soil, the chemistry of cadmium is largely controlled by pH. Cadmium may be adsorbed on clay minerals, carbonates or hydrous oxides of iron and manganese or may be precipitated as cadmium carbonate, hydroxide, & phosphate. Under acidic conditions, cadmium solubility increases, and very little adsorption of cadmium by soil colloids, hydrous oxides, and organic matter takes place.

Both toxicity and bioavailability of cadmium are influenced by soil characteristics. Cadmium mobility & bioavailability are higher in more acidic soils, and lower in chalky/ lime soils. One way to reduce cadmium bioavailability is to lime the soil to make it less acidic, However, once cadmium is in soil, it is persistent and cannot be broken down into less toxic substances in the environment.

Cadmium enters agricultural soils from the atmosphere & from application of phosphate fertilizers and sewage sludge. In heavily contaminated areas, re-suspension of dust can cause a substantial proportion of crop contamination and human exposure via inhalation & ingestion.

The presence of cadmium in fertilizers and atmospheric deposition had been found to cause increasing amounts of cadmium in topsoil in a number of European countries. If zinc is present, it can reduce cadmium's availability to plants, by inhibiting calcium uptake and preventing it from moving from the roots to the shoots of the plants.

Although cadmium emissions and concentration in the air have been reduced, data do not show reduced body burden of cadmium in non-smokers. In the top layers of arable soil, more cadmium is typically being deposited than is being removed: cadmium is accumulating in certain soils, increasing the likelihood of future exposure through food. Cadmium levels tends to increase moving up the food chain (bioaccumulation) .

There is evidence that elevated cadmium exposure during pregnancy may affect a child's motor skills & perception, and that high cadmium levels in the urine of school children are associated with a weakened immune system.

The health effects of low-level, chronic exposure to cadmium may be quite different to the high levels that caused itai itai disease. Exercises mapping the levels of cadmium in Europe suggest correlations between cadmium and age-adjusted prostate or breast cancer rates.

The main sources of cadmium directed to landfills and waste deposits are municipal waste, cadmium processing, non-ferrous metal processing & cement production, and both industrial & municipal wastes are important sources of cadmium for landfilling.

In densely-populated countries, such as Denmark & the Netherlands, waste incineration residues, and in particular clinker, are frequently used for road construction purposes and other civil works, thereby increasing the possibility that cadmium may be spread into environment through construction and, later, reconstruction activities

Lead

Exposure to lead during the early stages of children's developments is linked to a drop in intelligence. Studies suggest that for each 10 μ g/dl (microgram per decilitre) of blood lead, IQ is reduced by at least 1&3 points.

Phasing out lead from petrol has had an effect on levels of lead measured in children's blood. Soil lead and house dust, but not lead based paint, are associated with population blood lead levels in children.

Most soil lead and house dust are associated with leaded gasoline. Levels of lead in the blood began to decline in many countries largely because unleaded petrol was gradually introduced in these countries. Besides car exhausts, industrial emissions are important sources of exposure to lead.

In Poland, the geometric mean of lead levels in the blood of children living in the vicinity of zinc & copper mills is shown to have ranged between 7.4 and 11.4 μg/dl. This is in contrast to just 3.0–6.3 μg/dl among children living in five towns where there were no industrial lead emitters. In Hungary, the survey was conducted in areas previously contaminated with lead, where contamination had been due to either unlawful disassembly of used car batteries or run-off water from a land mine. The data shows evidence that there are still children with elevated blood lead levels.

Children are particularly at risk from adverse effects of lead exposure because, intake of lead per unit of body weight is higher for children than for adults. Young children often place objects in their mouths, resulting in the ingestion of dust & soil and, possibly, increased intake of lead. Physiological uptake rates of lead in children are higher than in adults. Young children are developing rapidly, their systems are not fully developed, and so they are more vulnerable than adults to the toxic effects of lead.

Lead in the environment has multiple sources, including petrol, industrial processes, paint, solder in canned foods, & water pipes. It can affect human health via a number of pathways, including air, household dust, street dirt , soil, water and food.

Lead-containing petrol has been a major source of lead pollution and is a significant contributor to the lead burden in the body in the countries where still used, most top soils in inhabited parts of the globe are enriched with lead. Industrial emissions are also important sources of lead contamination of the soil & ambient air, and lead may also be ingested from atmospheric air or flaked paint that has been deposited in soil & dust, raising blood lead levels. In addition, food and water may also be important media of baseline exposure to lead.

Mercury

Exposure to methyl mercury, the most harmful form of mercury to human health, affects brain development, resulting in a lesser IQ, and consequently a lower earning potential. Once methyl mercury is formed, it cycles through the environment for thousands of years, exposing humans and other species to potentially toxic levels for generations.

Large amounts of mainly inorganic mercury have accumulated in the environment, especially in soils & oceans, as a result of these past emissions and releases from human activities. Although mercury pollution can occur naturally in the environment through events such as forest fires, most comes from the burning of fossil fuels. The greatest harmful exposure to mercury for human is through eating fish (besides direct ingestion of contaminated soil by young children).

Cement production, mining & smelting, artisanal & small-scale gold mining. burning coal and oil refining are some of the activities emitting mercury which can build up in soils. Consumer products such as electronic devices, switches, batteries, energy-efficient light bulbs and certain cosmetics, dentistry, plastic production, and the chlor-alkali industry are also contributors to mercury emissions.

After it is deposited in soils & sediments, bacteria and microbes are mainly responsible for changing mercury to methyl mercury. Over 90% of the mercury found in fish is methylmercury. Mercury can enter the food chain via agricultural products or seafood. Mercury's use in agriculture has led to distressing human health incidents. At least 459 people died in Iraq when flour was made from grain treated with a fungicide containing mercury in 1971. Children whose mothers ate contaminated bread when they were pregnant were the worst affected. Agricultural products used today may still contain mercury. Rice crops grown in areas with high levels of coal powered industry, mining or smelting have also shown to be affected recently.

A team of Chinese investigated dietary mercury contamination in rural, inland China – a region where few people eat fish. They focused on Guizhou province, which has 12 large mercury mining and smelting operations, plus other heavy coal powered industry. The researchers looked at mercury levels in foods eaten by populations from several locations: a village located inside a nature preserve, a region downwind of a major coal plant, people living near a defunct zinc smelter and a community whose air was polluted by mercury mining operations. Mercury exposures for these communities varied considerably, but in every one of them 'rice accounted for 94-96% of the probable daily intake of methyl mercury'. One reason is that rice paddies here contain the types of bacteria that can convert inorganic mercury to its more toxic, methylated form.

Most mercury contamination sites are concentrated in industrial areas, but mercury can also travel long distances to locations far away from its production of use. Mercury levels in the atmosphere will fall fairly rapidly when emissions cease, but it will take many decades for levels in soils or oceans to fall.

The EU banned mercury exports in 2011. Under EU law, mercury that is no longer used by the chlor-alkali industry or that is produced in certain other industrial operations must be put into safe storage.

Modern humans, along with other mammal and bird species, have up to ten times the levels of mercury in their tissues as in pre-industrial times (before 1800). Researchers gather this data from samples including teeth, hair, feathers and egg shells.

The extent to which mercury is released from soils into waters depends on climate and topography. Many questions also remain about how mercury transfers around the whole ecosystem, including its uptake by living organisms and people.

Asbestos

Asbestos contamination in the soil is of concern in a number of locations, because it can be released to the air by the wind or by human disturbance. Asbestos has long-term health consequences if it is inhaled, with increased mortality from lung cancer and mesothelioma the most extreme outcomes.

Disturbing contaminated soil can pose an inhalation risk, and studies have shown asbestos concentrations of health concern may be released from that contain only low levels (less than 1%) asbestos contamination.

Researchers established links between asbestos exposure and lung malignancies, such as lung cancer & malignant mesothelioma (MM) and people living closes to mines or asbestos plants have been found to have an increased chance of contracting MM (in addition to workers at the plants, who have suffered occupational exposure). Clear cut cases have also been established of asbestos in soil causing health impacts for residents.

Dioxins and Dioxin-like chemicals

Dioxins are a group of chemically-related compounds that are persistent organic pollutants (POPs). Highly toxic, dioxins accumulate up the food chain, with the highest levels found in animals at the top of the food chain (an effect known as 'bio magnification'). More than 90% of human exposure to dioxin is through food, mainly meat & dairy products, fish and shellfish.

The chemical name for dioxin is: 2, 3. 7,8- tetrachlorodibenzo para dioxin (TCDD). The name dioxins' is often used for the family of structurally & chemically related polychlorinated dibenzo para dioxins (PCDDs) and polychlorinated dibenzofurans (PDFs). Certain dioxin-like polychlorinated biphenyls (PCBs) with similar toxic properties is sometimes labeled dioxins. Scientists have identified some 419 types of dioxin-related compounds, but only about 30 of these are considered to have significant toxicity, with TCDD being the most toxic.

Dioxins are highly toxic and can cause reproductive & developmental problems, damage the immune system, interfere with hormones and cause cancer. Dioxins are everywhere-all humans have background exposure leading to a body burden of these chemicals, which is not expected to affect human health when low. But due to the highly toxic potential of this class of compound, reducing current background exposure is advisable. Once dioxins have entered the body, they remain for a long time because of their chemical stability and their ability to be stroked in body fat.

Short-term exposure of humans to high levels of dioxins may result in skin lesions, such as chloracne & patchy darkening of the skin, and altered liver function. Long-term exposure is linked to impairment of the immune system. Chronic exposure of animals to dioxins has resulted in several types of cancer. TCDD is classified as a 'known human carcinogen'. However, it does not affect genetic material and there is a level of exposure below which cancer risk would be negligible.

The developing foetus is most sensitive to dioxin exposure. The newborn, with rapidly developing organ systems, may also be more vulnerable to certain effects. Some individuals or groups of individuals may be exposed to higher levels of dioxins through their diets (e.g. high consumers of fish in certain parts of the world) or their occupations eg, in paper industry, in incineration plants or at hazardous waste sites.

Since 1987, the WHO has conducted periodic studies on levels of dioxins in human milk. Levels of dioxins have decreased in Swedish breast milk. However, levels of the flame retardant chemicals polybrominated diphenyl ethers (PBDE) are increasing. Although manufacture and new use of penta & octa-PBDE formulations is banned in the EU & china, large quantities of the flame retardants remain in consumer & industrial goods and continue to enter waste streams.

The routes for dioxins to enter food are complex, and there are many sources. The largest source is past application of contaminated herbicides on agricultural soils. Waste incineration, industrial processes and deposition onto soils from atmospheric fallout are also significant sources, in addition to sewage sludge application.

When dioxins enter soils, they remain in the very top layer (the top 0.1 cm) with a half-life (time taken for concentration to halve) of 9-15 years. At deeper soil levels, dioxin can persist for 25-100 years. With dioxins persisting in the human body with a half-life of up to 11 years, it can be quite difficult to make direct correlations between concentrations of dioxins in human tissue and local soil.

Although dioxins form locally, their environmental distribution is global. The highest levels of these compounds are found in some soils, sediments and food, especially dairy products, meat, fish & shellfish. Very low levels are found in plants, water and air.

Extensive stores of PCB-based waste industrial oils, many with high levels of PCDFs, result in dioxin release into the environment and the contamination of human & animal food supplies. PCB-based waste is not easily disposed off without contamination of the environment and human populations. Such material needs to be treated as hazardous waste and is best destroyed by high temperature incineration.

Many countries monitor their food supply for dioxins. This has led to early detection of contamination and has often prevented impacts on a larger scale. In 1999, high levels of dioxins were found in poultry and eggs from Belgium, subsequently, dioxin-contaminated animal-based food (poultry, eggs, pork) were detected in several other countries. The cause was traced to animal feed contaminated with illegally disposed PCB based waste industrial oil.

Another case of dioxin contamination of food occurred in the US in 1997. Chickens, eggs, and catfish were contaminated with dioxins when a tainted ingredient (bentonite clay, sometimes called 'ball clay') was used in the manufacture of animal feed. The contaminated clay was traced to a bentonite mine. Investigators speculate that the source of dioxins may be natural, perhaps due to a prehistoric forest fire.

Large amounts of dioxins were released in serious accident at a chemical factory in Seveso, Italy, in 1976. A cloud of toxic chemicals, including TCDD, was released into the air and eventually contaminated an area of 15 square kilometers where 37,000 people lived. Extensive studies in the affected population are continuing to determine the long-term human health effects. Studies have shown increases in cancer, circulatory diseases, chronic obstructive pulmonary disease and diabetes following the incident. TCDD has also been extensively studied for health effects linked to its presence as a contaminant in some batches of the herbicide, which was used as a defoliant during the Vietnam War. A link to certain types of cancers and also to diabetes is still being investigated.

Organic pollutants, including hazardous pesticides

Organic (carbon-based) pollutants include pesticides. Those that were once released into air or water will end up in soils, with the exception of those that are deposited at the bottom of oceans. Among organic pollutants some are referred to as 'POPs' or persistent organic pollutants, which do not break down quickly in the environment.

Types of organic pollutants found in soil include:

- Polychlorinated biphenyls (PCBs)
- Polychlorinated dibenzofurans (PCDFs)
- Polycyclic aromatic hydrocarbons (PAHs)
- Organophosphorus and carbamate insecticides (pesticides)
- Herbicides
- Organic fuels (gasoline, diesel)
- Pharmaceuticals and their metabolites

Health effects of organic pollutants include the following

Individuals with increased exposure to pesticides (e.g. farmers, landscapers) appear to be at greater risk of non-Hodgkin lymphoma. Studies have also shown a greater risk among children from homes where pesticides are frequently used, or otherwise subject to higher exposure.

Some studies have linked leukemia, particularly in children, with insecticides. Timing of exposure is significant, with children exposed in the womb most likely to be affected. Pesticides have also been linked to brain cancer in children whose parents are exposed to high levels of pesticides, for example, through work.

- Some studies have linked pesticides to breast cancer and benign breast changes, although other studies have produced findings that disagree.
- Researchers have linked pesticide exposure to kidney cancer and pancreatic cancer.
- Several studies have linked pesticides to increased risks of prostate cancer, particularly reliant to the fumigant methyl bromide.
- Links have also been made between stomach and atrazine.
- Long-term effects of pesticides on the nervous system include cognitive & psychomotor dysfunction, and neurodegenerative & neurodevelopmental effects. Pesticide poisonings result in well described acute & chronic neurotoxic syndromes. Chronic effects from low or moderate exposures have been less well documented.
- Many studies have shown that occupational pesticide exposure could increase the risk of later developing Parkinson's disease.
- Studies have consistently shown an increased risk of birth defects resulting from parental exposure to pesticides. The majority of studies looking at the effects of pesticides on foetal growth showed that agricultural pesticides altered foetal growth, and increased risks of miscarriage.
- Those exposed to pesticides also had a greater frequency of chromosome aberrations (genotoxic effects), although it is hard to separate these effects from other sources of genetic damage, such as smoking, alcohol or radiation.

However, we are seldom exposed to just one of these chemicals or in high doses-most people are exposed to a complex mixture of these chemicals at low concentrations.

Overall, few studies have been conducted on the toxicity of complex chemical mixtures in soils. The effects of the soil and organisms within it upon organic pollutants are unknown. The data that do exist tend to be on short-term, high level exposure of these chemicals, which is less relevant to the potential low-level, long term health impacts from living near to contaminated soil.

Synthetic organic chemicals often include atoms (such as chlorine, bromine or sulfur) that have been inserted into their structures in positions not commonly found in nature. This is one of the factors that makes synthetic organic chemicals hard for natural processes to deal with-they are toxic to living creatures and do not break down easily, as well as persisting in the environment. This also means that they can build up over time in body tissues, or become magnified along food chains.

The chemical industry creates (or previously created) these compounds in very large quantities, to use in a wide range of products including plastics, refrigerants, preservatives and pesticides. Many POPs are actually by-products of these processes, and are not useful in their own right. Examples include dioxins, and polychlorinated dibenzofurans (PCDFs)

Difficulties in measuring soil contamination and health impacts

There are many methodologies involved around soil sampling, background exposure levels and health study.

Knowledge of the level of a contaminant in soil and human susceptibility in its presence does not give an accurate picture of the toxicological risk to humans, and only indicates risks originating from the soil. To get an idea of the true public health risk, other sources that people are exposed to alongside soils must also be taken into account. This 'background intake' may come from food, drinking water, or air pollution levels. Individual behavior, such as smoking, unusual diet or occupational exposure may also contribute to the overall impact on health from soil contaminants.

Levels of contaminant vary depending on soil depth. Whether the soil has been moved or landscaped will also have an impact. Standardized sampling methods are important as a site will not necessarily have a uniform distribution of contaminants throughout its soils, so the levels, and risks, will vary from one area to the next. In cases of sampling at a large plant or facility, accurate measurements of contamination levels at the site may not be possible until the plant has been decommissioned and the site cleared.

The behavior of contaminants in soils depends on several physical & chemical processes:

- Reduction/oxidation (redox)
- Absorption
- Precipitation
- Desorption

Whether a soil absorbs or releases a pollutant depends on:

- Types of mineral present
- Amount of organic material present
- Soil pH (acidity)
- Redox potential
- Moisture

How effectively contaminant molecules bond with the surface of the soil (absorbs) is important when considering the human health risks. A strongly-bonded contaminant will be less likely to leach out of the soil into groundwater, or be released from the soil as a vapour than one that is weakly bonded. These properties relate to the contaminant's bioavailability.

In contrast to their behaviour in air and water, pollutants in the soil do not generally disperse quickly. They often form discrete pockets of pollution (hotspots). Soils are made up of solid liquid and vapour phases. How the pollutant is divided between these phases will determine how it behaves, and what types of environmental & public health risk will result.

Estimates of soil toxicity depend on the techniques used to measure them, and the interpretation of the results. Scientists working in land contamination risk assessment frequently use in vitro (laboratory only) tests to estimate bio accessibility of contaminants from the soils on their sites. Most risk assessment models use exposure estimates based on intake, rather than uptake. 'Intake' is defined as the amount of a substance an individual is exposed to (in weight per kg of body weight per day), and 'uptake' as amount of a substance taken up by the body that enters the bloodstream (in µg/100ml).

Bio accessibility: The fraction of a substance that is released from soil during processes like digestion into solution (the so-called bio accessible fraction) .

Bioavailability: The fraction of the chemical that can be absorbed by the body through the gastrointestinal system, the pulmonary system and the skin. A soil guideline value (SGV) is a figure for the concentration of a contaminant in the soil that sets off 'possible risk' alarm bells. It means that further investigation and/or risk management are needed. These SGVs are generally derived from estimates of toxicity from a certain human intake of the soil rather than actual human uptake of the contaminant. For example, a child may ingest some soil containing a quantity of a heavy metal known to be above risk levels for negative health impacts, which is classed as the intake. But the percentage of that heavy metal that is actually taken in and processed by the child's body is the uptake.

As soil properties may vary considerably from one site to another, even soils contain the same levels of a certain contaminant, the bioavailability of that contaminant to humans may be quite different, depending, for example, on how tightly the chemical is bound to the soil.

Two different definitions of bioavailability are used in human health assessment:

Absolute bioavailability

The percentage of an external chemical dose that reaches the bloodstream (the ratio of internal dose to the administered dose).

Relative bioavailability

This compares the absolute bioavailability of different forms of a contaminant, or exposure to different media containing a contaminant.

This is important for soil studies, because the chemical form that the contaminant takes in the actual samples of soil in question, as well as the makeup of the soil itself, could differ from the 'reference' sample used to derive the established risk limit values for that contaminant in soil.

Some soil contaminants can be extremely long-lived in the environment, as noted. The complexity of calculating not just the current risk from a site, but also its future risks, is considerable. A great deal depends upon the stability of the chemical contaminants in the soil over time, and upon the site's future uses.

5 CHAPTER

Prevention & Control of Land/Soil Pollution

There has been an increase in the natural calamities, all over the world, in the past few years. One of the main reasons behind this is the environment degradation that has taken place due to air, water and land pollution. Speaking of land pollution, the main culprits behind it are the industrial wastes, which are not disposed off properly. These accumulate overtime, seeping into the soil & land, adversely affecting, not only humans, but plants and animals as well.

Prevention of land pollution can be done by:

- **Proper Disposal of Wastes**

  Proper wastes management is the most important things to do for the prevention of land pollution. Several steps should be taken to ensure that the wastes get disposed in the most harmless way, i.e. produces less of toxins. One of the ways of doing this is by separating dry and wet waste. Dry wastes should be recycled or it can be burned. On the other hand, wet waste should be decomposed. This decomposed wastes should be further dumped in pits so that it can be converted into compost and used in agriculture. When disposing off waste, it should be treated with various kinds of chemicals, so that it has a very low level of toxicity.

- **Recycling and Reuse**

  People should buy and use products that have been recycled. For instance, plastic bags should be given up, instead, jute bags should be used. Reusing old items is another good way to reduce the stress on various resources. Clothes, paper, glass, everything should be recycled and reused, as far as possible.

- **Use of Biodegradable Products**

  People should buy biodegradable products. Go organic- that's the mantra for avoiding land pollution. Organic farming, lessening the use of pesticides, buying organic fruits & vegetables and other products, will keep land pollution in check.

- **Grow More Trees**

  Initiative should be taken to grow more trees. Trees can actually fertilize the soil. This means that by growing tress, soil will become fertile, agriculture will get boosted and land pollution will greatly reduce.

- **Efficient Use of Resources**

  There should be minimal wastage. There should be no littering. Oil and petrol should be used judiciously. So should all the other products and resources made from glass, plastic, metal & paper.

Both individuals and authorities have to make positive changes in the way they control things. Spreading awareness amongst the common people, use of organic, biodegradable products and to efficiently use resources, should be taken up seriously by the government. The industries too should follow proper procedures of industrial waste disposal. In addition to this, trees should be saved and more plants should be grown. All these initiatives are mandatory for any project on the prevention of land pollution, to be successful.

Land pollution, which is otherwise known as soil pollution, can be defined as the degradation of soil as well as land surfaces, due to various reasons. The most common causes of land pollution is solid waste that is dumped in large quantities. Others include the chemicals used for agricultural and industrial waste. Such pollution is a health hazard for both humans as well other living beings on the earth. In the long run, land pollution may disrupt the balance of nature, thereby, causing natural calamities.

Land pollution is mainly caused by solid waste that is dumped on the Earth. There are various factors that are responsible for the increasing amount of waste that has been accumulating on the land. This includes the rise in population, rapid urbanization and industrialization. Even lack of awareness about land pollution causes & effects and neglecting pollution control norms are certain factors that contribute to this condition. It has been observed that waste disposal is not done properly. While some of the biodegradable wastes take long to decompose, the non-biodegradables ones, like plastic and electronic goods remain the same for years, posing a health threat.

Even the industrial wastes and chemicals can cause degradation of the soil as well as water. This also applies to the fertilizers and pesticides used for agriculture. So there are many different causes of land pollution. Apart from being a health concern, land pollution will also affect the flora and fauna detrimentally. While some of these waste materials (dumped on land) release harmful chemicals that can pollute the soil as well as groundwater, they may also hinder the process of groundwater accumulation. Eventually, it may also contribute to natural calamities.

Solution for Land pollution

The following are some of the solutions for land pollution:

- As the most common causes for land pollutions is waste disposal, most of the control measures are associated with that. Proper wastes disposal is one of the golden rules for curbing land pollution. This is much more important in case of disposal of toxic wastes products. Industries must follow the laws with regard to waste disposal.
- One of the major land pollution solutions is recycling. Apart from reducing the amount of wastes products in the landfills, it also helps in curbing the dumping of non-biodegradable wastes on earth. Recycling is also beneficial in lowering other forms of pollution, cost savings and savings of energy resources.
- One must also try to reuse materials, if possible. Reusing is always beneficial than buying new ones. For example, you may use plastic and cardboard containers for further use. Otherwise store them and sell them for recycling. You also make some money by refraining from buying new products and also by selling the unwanted materials for recycling.
- So, each and every individual can contribute for this good cause, by making efforts for reducing land pollution. We should reduce the use of non-biodegradable materials. For examples, we may carry paper or cloth bags with us so as to avoid taking a plastic carry

bag from the shop. Switch over to the use of biodegradable materials. Use glass or metal products instead of plastic ones.

How to Reduce Land Pollution

Following steps can be taken to stop or reduce or avoid land pollution:

- Adoption of organic farming instead of using synthetic chemicals.
- Adoption of reclamation, recycle and reuse of waste materials.
- Proper disposal of industrial wastes into secured landfill sites.
- Educating people not to litter.
- Proper collection, transportation and disposal of municipal wastes.
- Reducing use of herbicides & pesticides.
- Organic waste matter should be disposed off in areas that are far away from residential places.

CONTROL OF SOIL POLLUTION

Soil may be polluted and converted into acidic soil or alkaline. It should be corrected by suitable technology, before cultivation.

Method of Soil treatment

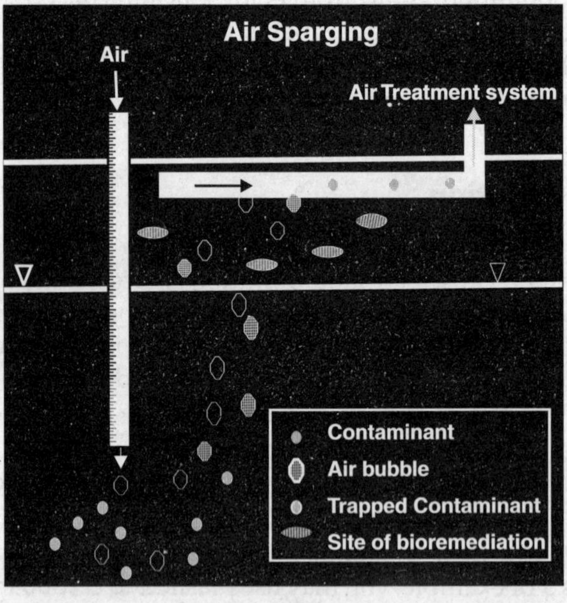

Fig.5.1: Air Sparging as a Method of Soil Treatment

Air sparging is an in situ remedial technology that reduces concentration of volatile constituent in petroleum products that are absorbed to soil and dissolved in groundwater. This technology, which is also known as "in situ air stripping" and "in situ volatilization, "involves the injection of contaminant-free air into the subsurface saturated zone, enabling a phase transfer of hydrocarbons from a dissolved state to a vapor phase. The air is then vented through the unsaturated zone. Air sparging is most often used together with soil vapor extraction (SEV), but it can also be used with other remedial technologies.

Soil washing is a water-based process for scrubbing soils ex situ to remove contaminants. The process removes contaminants from soils in one of the following two ways:

- By dissolving or suspending them in the wash solution (which can be sustained by chemical manipulation of pH for a period of time); or
- By concentrating them into a smaller volume of soil through particle size separation, gravity separation, and attrition scrubbing.

The concept of reducing soil contamination through the use of particle size separation is based on the finding that most organic & inorganic soil particles adhere to the soil. The silt & clay, in turn, are attached to sand & gravel particles by physical processes, primarily compaction and adhesion. Washing processes that separate the fine (small) clay & silt particles from the coarser sand & gravel soil particles effectively separate and concentrate the contaminants into a smaller volume of soil that can be further treated or disposed off . Gravity separation is effective for removing high or low specific gravity particles such as heavy metal-containing compounds (lead, radium oxide, etc.). Attrition scrubbing removes adherent contaminant films from coarser particles. However, attrition washing can increase the fines in soils processed. The clean, larger fraction can be returned to the site for continued use. Soil washing generally considers a media transfer technology. The contaminated water generated from soil washing are treated with the technology(s) suitable for the contaminants. The duration of soil washing is typically short to medium term.

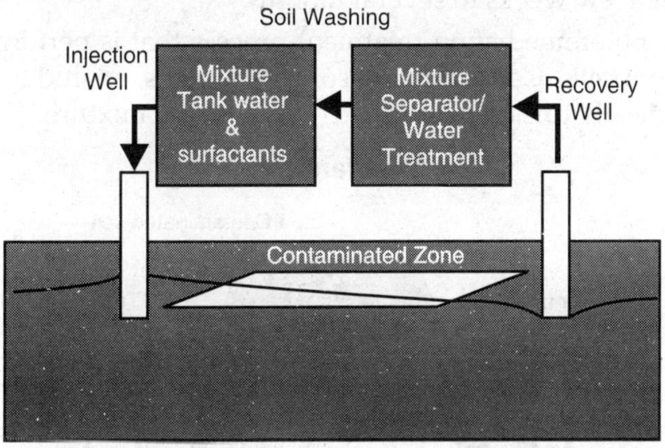

Fig.5.2: Soil Washing as a Method of Soil Treatment

Bio pile treatment is a technology in which excavated soils are mixed with soil amendments and placed on a treatment area that includes leachate collection. It is used to reduce concentration of petroleum constituents in excavated soils through the use of biodegradation. Moisture, heat, nutrients, oxygen, and pH can be controlled to enhance biodegradation.

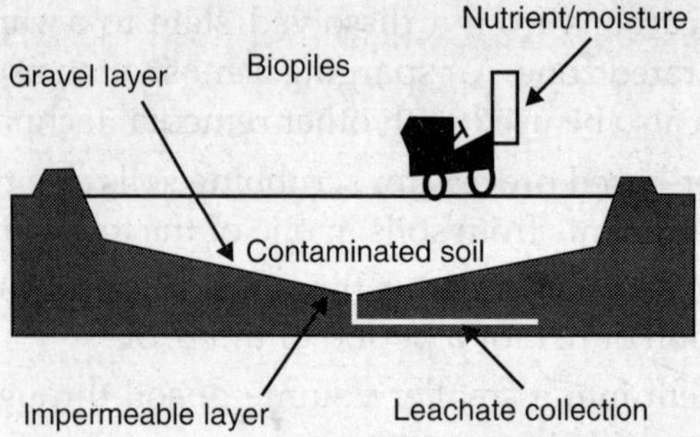

Fig.5.3: Bio pile soil treatment

The treatment area will generally be covered or contained with an impermeable liner to minimize the risk of contaminants leaching into uncontaminated soil. The drainage itself may be treated in a bioreactor before recycling. Vendors have developed proprietary nutrient and additive formulations & methods for incorporating the formulation into the soil to stimulate biodegradation.

Soil piles and cells commonly have an air distribution system buried under the soil to pass air through the soil either by vacuum or by positive pressure. The soil piles in this case can be up to 6 meters high (generally recommended, 2-3 meters maximum). Soil piles may be covered with plastic to control runoff, evaporation, & volatilization and to promote solar heating. If there are VOCs (Volatile Organic Chemicals) in the soil that will volatilize into the air stream, the air leaving the soil may be treated to remove or destroy the VOCs before they are discharged to the atmosphere. It is a short-term technology. Duration of operation and maintenance may last a few weeks to several months.

Land Farming is a bioremediation treatment process that is performed in the upper soil zone or in bio treatment cells. Contaminated soils, sediments, or sludge incorporated into the soil surface are periodically turned over (tilled) to aerate the mixture.

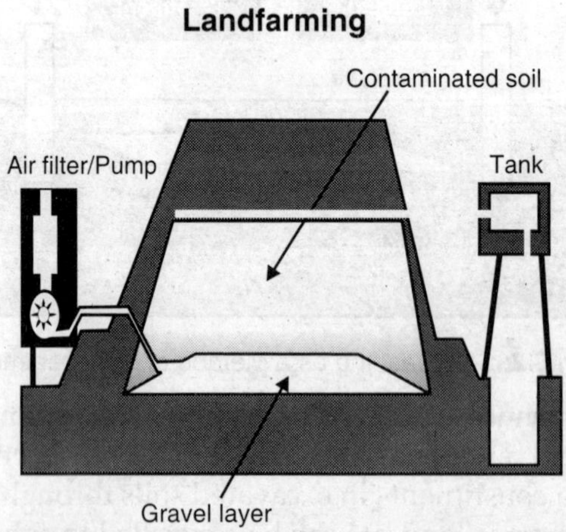

Fig.5.4: Land farming as a method of soil treatment

This technique has been successfully used for years in the management and disposal of oily sludge & other petroleum refinery wastes. In situ systems have been used to treat near surface soil contaminations for hydrocarbons and pesticides. The equipment employed in land farming is typical of that used in agricultural operations. These land farming activities cultivate and enhance microbial degradation of hazardous compounds.

BIOREMEDIATION

This means the use of naturally occurring microorganisms such as bacteria, fungi & plants to break down or degrade toxic chemicals compounds that have accumulated in the environment.

It is a method that treats the soil and renders them non-hazardous, thus eliminating any future liability that may result from landfill problems or violations.

Mechanism of bioremediation for soil pollution control is shown below in figure.

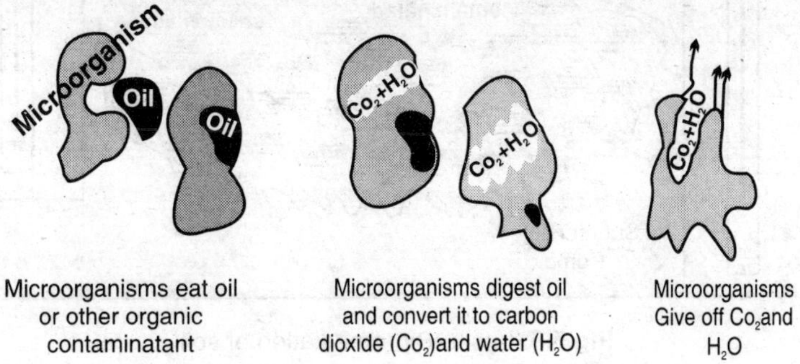

Fig.5.5: Bioremediation as a method of soil treatment

Factors affecting bioremediation are:

- Microbial factors.
- Temperature favorable for organisms.
- Availability of water (Moisture content)
- Availability of nutrients (N,P,K)
- C: N (carbon: nitrogen) ratio of the contaminant material (< 30:1)
- pH
- Availability of Oxygen in sufficient quantity in soil.

Types of bioremediation:

(a) **In situ bioremediation:** The soil treatment in place without excavation of contaminated soils or sediments.

(b) **Ex situ bioremediation:** It requires pumping of the groundwater or excavation of contaminated soil prior to remediation treatments.

Types of In situ Bioremediation:

- **Bio stimulation:** To stimulate the activity of microorganisms by adding nutrients and electron acceptors (e.g. O_2)
- **Bioventing:** injecting air through soil to stimulate microbe growth in unsaturated zone.
- **Bio augmentation:** Inoculation of soil with microbes or adding exogenous microbes to the subsurface.

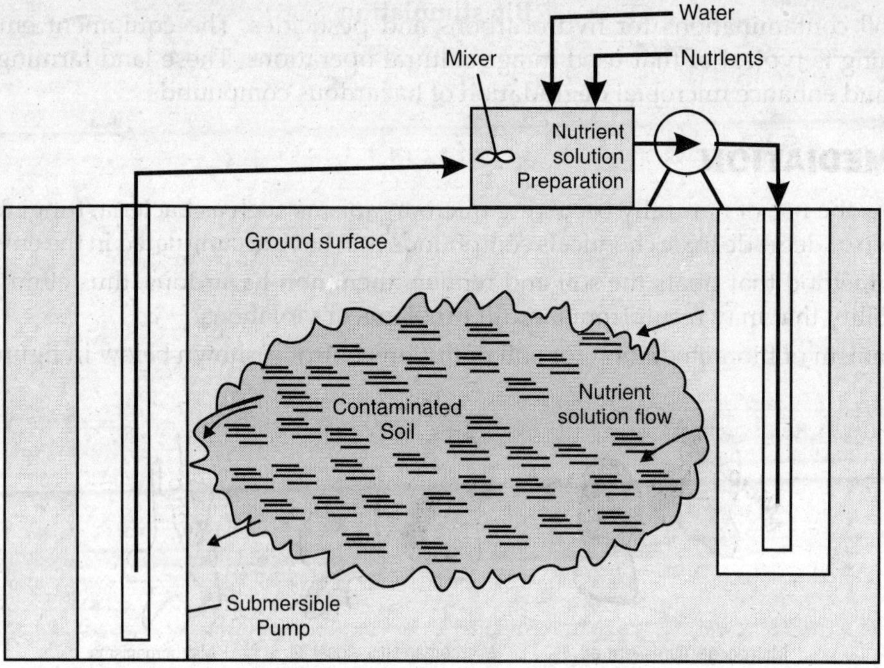

Fig.5.6: In-situ-Bioremediation of soil

Features of In-Situ-Bioremediation

- Good for large volumes
- Slower
- Doesn't work well in clays or highly layered sub surfaces
- Bio stimulation (stimulates biological activity)
  - Bioventing (Inject air/nutrients into unsaturated zone-good for mid weight petroleum, jet fuel)
  - Biosparging (inject air/nutrients into unsaturated and saturated zones)
- Bio augmentation (inoculates soil with microbes)
- Less expensive
- Creates less dust
- Less possibility of contaminant release into environment

Features of Ex-situ- Bioremediation

- Easier to control
- Uses to treat winder range of contaminants and soil types
- Costly
- Faster

Bio stimulation

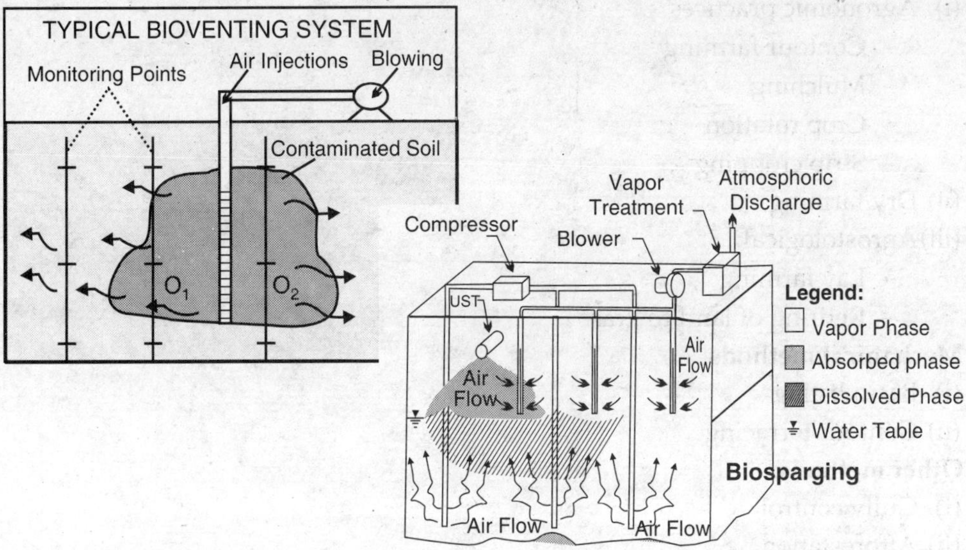

Fig.5.7: Bio stimulation techniques for stimulating biological activity

Types of ex-situ bioremediation:

(a) **Slurry-phases ex. situ bioremediation**-Soil combined with water/additives in tank, microorganisms, nutrients, oxygen added.

(b) **Solid-phases ex-situ bioremediation**:

- **Land-farming:** soil put on pad, leachate collected.
- **Soil bio piles:** soil heaped, air added.
- **Composting:** biodegradable wastes mixed with bulking agent.
- **Land Applied:** wastes added directly to soil which is later planted to a crop.

Advantages of Using Bioremediation Processes Compared with Other Remediation Technologies

- Biologically-based remediation detoxifies hazardous substances instead of merely transferring contaminants from on environmental medium to another;
- Bioremediation is generally less disruptive to the environmental than excavation-based processes; and
- The cost of treating a hazardous waste site using bioremediation technologies can be considerably lower than that for conventional treatment methods: vacuuming, absorbing, burning, dispersing, or moving the material.

Soil conservation

Soil conservation is the protection of soil against excessive loss of fertility by natural, chemical, or artificial means. It encompasses all management and land-use methods protecting soil against degradation, focusing on damage by erosion & chemicals. Soil conservation techniques can be achieved through crop selection & rotation, fertilizer & lime applications, tilth, residue management, contouring & cropping, and mechanical methods (e.g., terracing).

- **Biological methods:**
  (i) Agronomic practices:
    - Contour farming
    - Mulching
    - Crop rotation
    - Strip cropping

  (ii) Dry farming
  (iii) Agrostological:
    - Lay farming
    - Retiring of land to grass
- **Mechanical methods:**
  (i) Basin listing
  (ii) Contour terracing
- **Other methods:**
  (i) Gully control
  (ii) Afforestation

Soil Amelioration

(A) Soil Amelioration

Amelioration of Acidic Soil: Soil acidity is due to the accumulation of H^+ ions over H^- ions. Limiting material – are neutralization of H^+ ions such as:

- ***Quicklime***- oxide of lime usually known as burned lime or quicklime.
- ***Slaked lime***- can be obtained by adding water to quick lime.
- ***Blast furnace slag***- a byproduct during the manufacturer of pig iron viz, calcium silicate.
- ***Basic Slag***- is a byproduct of the basic open hearth methods of producing steel from pig iron,
- ***Electric Furnace slag***- is produced from the electric furnace reduction of phosphate rock during preparation of phosphorous. The products is mainly the calcium silicate.

The other methods which could result in amelioration of acidic soil are:

- Use of basic fertilizers such as ammonium nitrate reduces the soil acidity.
- Proper soil and water management.
- Usage of coral shell, chalk, wood ash, press mud, byproduct material of paper mills, sugar factories, fly ash and sludge etc.

Amelioration of Saline and Alkali Soil

Saline soil contains an excess of soluble salts. Saline soil reclamation can be achieved by:

- Providing proper drainage.
- Using salt free irrigation water.
- Use of acidic fertilizers-such as ammonium sulphate.
- Use of organic fertilizers.
- Use of organic manures.

Alkaline soil contains appreciable amounts of soluble salts. Alkali soil reclamation may be achieved by the following practices:

- Application of gypsum
- Use of sulphur
- Addition of organic matter.
- Addition of molasses.

(B) Prevention of solid waste dumping

Open dumping of solid wastes should be segregated and recyclable materials could be recycled. Other garbage can be converted into organic manure by suitable technology.

(C) Usage of bio-fertilizers and bio-pesticides.

(D) Following the concept of Integrated Plant Nutrient System (IPNS).

Organic / Sustainable Agriculture

Organic farming aims for the production of quality and safe agriculture products for consumption. This system requires less financial & external inputs and provides sustainable income to the farming community. Organic farming aims at production of quality & safe agriculture products which contain to chemical residues due to the adoption of eco-friendly production methods & farming system that restore and maintains soil fertility.

Organic farming is a production method which does not pollute the soil & ground water with chemicals residues and provides safe & quality food for consumption. It also increases the biological diversity of plants and animals that helps to maintain the natural eco balance. This approach also aims to recycle only the natural resources and restricts the use of external inputs which indirectly helps to reduce the energy consumption in the farming system considerably.

The vision of organic farming in India has necessitated the government to launch the National programme for organic production (NPOP) during 2000. By National accreditation policy & programme, the government has also implemented the National standards for various organic farming activities. Hence organic farming has to be promoted in a big way to provide quality & safe food to the growing population and also to protect the environmental degradation.

Organic farming aspires to a complex mix of agronomic, environmental, agricultural & processing and are based on a number of principles. They are:

- To produce food of high quality and safety.
- To interact in a constructive and life-enhancing way with natural systems and cycle.
- To consider the wider social & ecological impact of the organic production and processing system.
- To encourage and enhance biological cycles within the farming systems, involving microorganisms, soil, flora & fauna, plants and animals.
- To develop a valuable and sustainable aquatic ecosystem.
- To maintain and increase the long term fertility of soils.
- To promote the healthy use and proper care of water, water resources, & all life therein.
- To use, as far as possible, renewable resources in locally organized production systems.
- To create a harmonious balance between crop production and animal husbandry.
- To minimize all forms of pollution.
- To process organic products using renewable resources.
- To produce fully biodegradable organic products.

These principles are given equal importance as that of other economically viable production technologies.

Organic Farming Requirements

Achieving the above mentioned principles of organic farming needs a holistic farming system with integrated approach in all aspects. The basic principle of organic farming in enhancing the soil fertility can be achieved through proper recycling of organic wastes, versatile crop rotation & cropping systems, a wide range of biological methods for control of pests, diseases & weeds and to avoid the use of synthetic fertilizers, chemicals pesticides & herbicides. Habitat development is the key factor in restoring the natural eco-system which in turn facilitates the symbiotic co-existences of fauna and flora apart from promoting natural predators, parasites etc.

(a) Maintaining soil fertility

Depletion of soil organic matter under intensive cropping system is the key factors in altering biological equilibrium of the soil ecosystem. It is essential to maintain the soil food web, where all the soil organisms viz, bacteria, fungi, actioncetes, protozoa, earthworms etc., flourish in population in the presence of sufficient amount of soil organic matter. In order to maintain the soil fertility, the following farming practices are recommended.

- Increased use of organic manures, green manures.
- Enriched vermin compost and bio composts.
- Use of bio fertilizers.
- Crop rotation with high and low biomass crops.
- Avoiding the use of chemical fertilizers.

(b) Plant Protection methods

Indiscriminate use of chemicals, pesticides and herbicides leads to soil & ground water contamination which causes health problems in living systems. The accumulation of toxic residues in the food products has created considerable awareness among the producers and consumers. The reports on the pesticides residue in food products revealed that, most of the food products from conventional agriculture contain more than 70 per cent residues. In addition, it also impairs the soil microflora that is essential to maintain soil fertility. These problems can be solved by adopting organic farming practices which use only the natural bio pesticides for plant protection. Generally, bio pesticides, bio control agents, plant extracts etc. are used for controlling the pest and disease problems.

(c) Animal husbandry

The basis for including animal husbandry in the systems is to respect the physiological and ecological needs of animals. This is achieved by providing sufficient quantities of good quality organic fodder, shelters according to their behavioral needs and also by proper veterinary treatment. Animals are an important part of organic system because they act as the agents for recycling of byproducts with value addition. Further, they contribute to complete the nutrient cycle and maintaining soil fertility. They also contribute to draught energy for agricultural operations and provide essential manure for soil nutrition & urine for pesticides.

(d) Processing of organic products

The basis of processing organic products in that as far as possible the vital qualities of the products are maintained throughout each step of the process. This is achieved by choosing & developing methods which are adequate to the specification of the ingredients and by developing standards which emphasize careful processing methods, limited refining, energy saving technologies, minimal use of additives & processing aids etc. The production and handling of organic products in a safe way can be achieved by adopting existing standards or by developing new standards, which define the safe methods of wastes management in the form of products besides packing systems and energy saving systems in processing & transport.

Information needed to clean up contaminating materials added to soil:

- **Kind of material**-organic or inorganic; is the material biodegradable/ dangerous to animals & humans?
- **How much material** was added to the soil; will it overload the organisms in the soil?
- **C: N ratio** of the pollutant material.
- **Nature of soil:** Will the soil be able to handle the material before groundwater is contaminated?
- **Growing conditions for the soil organisms:** Is it cold, too wet etc.?
- **How long the material has been on site:** Is there evidence of environmental problems, is it undergoing decomposition?
- **Immediate danger to people & environment:** Urgency of the situation.

Control measure for soil pollution can be summarized as follow:

(1) Controlling soil erosion: Soil erosion can be controlled by a variety of forestry and farm practices. **Ex:**

- Planting tree on barren slopes.
- Contour cultivation and strip cropping may be practiced instead of shifting cultivation.
- Terracing and building diversion channels may be undertaken.

Reducing deforestation and substituting chemical manures by animal wastes also helps arrest soil erosion in the long term.

(2) Proper dumping of unwanted material: Excess wastes by man and animals pose disposal problem. Open dumping is the most commonly practiced technique. Now a days, controlled tipping is followed for solid waste disposal. The surface so obtained is used for housing or sports field.

(3) Production of natural fertilizers: Bio-pesticides should be used in place of toxic chemical pesticides. Organic fertilizers should be used in place of synthesized chemical fertilizers. Ex.: Organic wastes in animal's dung may be used to prepare compost manure instead of throwing them wastefully and polluting the soil.

(4) Proper hygienic condition: People should be trained regarding sanitary habits. **Ex:** Lavatories should be equipped with quick and effective disposal methods.

(5) Public awareness: Informal and formal public awareness programs should be imparted to educate people on health hazards by environmental education. **Ex.** Mass media, Educational institutions and voluntary agencies can achieve this.

(6) **Recycling and Reuse of wastes:** To minimize soil pollution, the wastes such as paper, plastic, metal, glasses, organics, petroleum products and industrial effluents etc. should be recycled & reused. **Ex.** Industrial wastes should be properly treated at sources. Integrated waste treatment methods should be adopted.

(7) **Ban on Toxic chemicals:** Ban should be imposed on chemicals & pesticides like DDT, BHC, etc. which are fatal to plants and animals. Nuclear explosions and improper disposal of radioactive wastes should be banned.

(8) Use of pesticides and fertilizers should be minimized.

(9) Cropping techniques should be improved to prevent growth of weeds.

(10) Special pits should be selected for dumping wastes.

(11) Controlled grazing and forest management.

(12) Wind breaks and wind shield in areas exposed to wind erosion.

(13) Afforestation and reforestation.

(14) Practice 3 R's: reduce, reuse, recycle.

APPENDIX 5.1

Control techniques for specific soil pollutants

A number of ways been suggested to curb the soil pollution rate. Some the steps to reduce soil pollution are:

- Ban on use of plastic bags below 20 microns thickness.
- Recycling of plastic wastes.
- Ban on deforestation.
- Encouraging social and agro forestry programmes.
- Reducing the use of chemical fertilizers and pesticides.
- Recycling paper, plastics and other materials.
- Ban on use of plastics bags, which are a major cause of pollution.
- Reusing materials.
- Avoiding deforestation and promoting forestation.
- Suitable and safe disposal of wastes including nuclear wastes.
- Chemicals fertilizers & pesticides should be replaced by organic fertilizers & pesticides.

Prevention of soil pollution

Toxic chemicals compounds, salts, radioactive agents, toxins and other wastes contribute to soil pollution. These have adverse effects on plant and animals health. Soil contains both organic as well as inorganic material. The organic material is formed due to decaying of plant and animal matter. This makes up the upper most layer of soil. The organic soil such as rocks, has taken over thousands of years to be formed. The top layers is made up of organic soil, while the layers below are inorganic layers as well. There are different types of soil pollution, namely agricultural soil pollution and industrial waste causing soil pollution. These different types of pollution cause the fertility of the soil to reduce and mineral content in the soil to be destroyed.

To increase agricultural yield, farmers use chemical fertilizers but it costs the soil losing its fertility. To restore the fertility of the soil, will take a very long time. Farmers should start using bio fertilizers. The microorganisms in these fertilizers will help increasing the fertility of the soil.

To avoid soil pollution, it is important, that along with fertilizers, farmers should shift to bio pesticides and fungicides, also known as herbicides. These products will take a little longer to reach, but they do not have adverse effects on the soil. It is best to use manure both as a fertilizer as well as pesticide, as it has far less side effects as opposed to chemical fertilizers & pesticides.

Toxic waste has a big role to play in soil pollution. Hence, industrial toxic waste should be treated to reduce its toxicity before it is disposed off. At the same time, responsible methods should be used for disposing off the waste. The best, however, is to avoid the use of harmful chemicals unless they are of extreme importance.

If each family recycles wastes, the land pollution caused due to landfills will be reduced considerably. The land so saved can be used constructively for a number of better tasks.

Plastic is one of the main causes of soil pollutions, as it takes a very long time to disintegrate. Therefore, people should consider shifting to reusable containers like glass, cotton bags, etc. Although paper does disintegrate faster, a lot of tress are cut for producing paper bags. Therefore, it is best to opt for cloth bags. Similarly, instead of using tissue paper in the kitchen, etc., one should opt using cloth napkins, handkerchief, etc. This will go a long way in reducing landfills.

Though the organic products are costly as opposed to the chemically grown products, its biodegrability will encourage more organic production. This will help in preventing soil pollution.

To prevent soil pollution, measures against deforestation has to be undertaken. Soil erosion is caused, when there are no trees to prevent the top layer of the soil from being transported by different agents of nature like water and air. At the same time, measure should be taken to avoid over cropping & over grazing, as it leads to flood & soil erosion and further deterioration of the soil layer.

Various chemicals such as pesticides, insecticides and fungicides play important roles in plant growth; however the overuse of these chemicals is considered to be one of the prime factors leading to soil pollution. Reducing or even avoiding the use of such chemicals is one of the most elementary and important preventive measures. Manures and bio-fertilization can be used as alternatives to chemicals. Use of manures are recommended because, as compared to pesticides, manures have minimal adverse effects on the environment. Bio-fertilization is a process in which certain microorganisms such as rhizobium are used to increase the soil fertility in places of chemicals.

Another common measure used to minimize soil pollution is controlling the growth of weeds. Weeds are unwanted plants that grow alongside the main plant and often result in the accumulation of various minerals into the soil layer. One of the common methods to control weed growth is covering the soil with layers of newspaper or plastic sheet just before cultivation.

Other common methods of preventing soil pollution includes reforestation and recycling of waste materials. Deforestation or the cutting down of tress leads to erosion of the soil, which leads to soil pollution due to the loss of fertility of the soil. Thus, reforestation is an effective method of preventing soil pollution.

In addition, reducing the volume of refuse or waste in landfills by recycling materials such as plastics, papers & various other materials is another effective and common method of preventing the phenomenon of soil pollution.

- Strong regulatory programs to minimize soil contamination need to be introduced.
- Reuse and recycle unwanted items. Or even better, reduce consumption and reduce your trash. The less rubbish we create, the less chance the wastes will end up in our soil.
- There is a need to educate the public about the harms done when they litter.
- For gardens, make use of organic fertilizers & organic pesticides, because they are usually made of natural substance, are bio-degradable and do little harm to the natural balance in the soil.
- Insist on buying natural and organic food, because chemicals pesticides & fertilizers are not used in their growing process. Not only are organic foods healthier for the environment, they are also healthier for us.

- Cut down uses of paper. Or use recycled paper. In this way, fewer trees need to be cut down and three would be reduced deforestation.
- For decades people have believed that harmful chemical pesticides were the only true way to rid gardens and crop fields from pests. Soil pollution has occurred from the use of pesticides and it takes years and sometimes decades for some of these chemicals to break down. Luckily there are many organic chemicals that are just as effective. The effects of the pesticides on soil micro-organisms are less invasive when organics pesticides are used. Pesticides need to break down quickly in the sunlight and in the soil. The faster chemicals breakdown, the sooner that soil can return to a healthy state. Most organic pesticides are also safe to use around people and pets. They can easily be washed from fruits and vegetables making them healthier for us to eat.

Soil Pollution Control Techniques

Soils, besides supplying food, also have purification property. This soil property is caused by their physical properties (water permeability operation from pores), their chemical properties (surface absorption) and their biological properties (decomposition of organic matters).

(a) Controlling Oil Pollutions in Soil

Oil and chemical materials & their derivatives may cause soil pollution as a result of transport or storage. If more oil materials penetrate into the more depth of soil, removing its pollution will be difficult. Some bacteria and microorganisms in soil can cause decomposition of oil materials.

Following are regarded as the ways to control effects of oil pollution:

- Preventing oil from spreading widely,
- Improve the soil ventilation through ploughing and mixing,
- Increasing food nutrients to the soil like nitrogen and phosphorous,
- Combining soil with microorganisms which decompose oil materials

(b) Controlling pollution Caused by Wastes in Soil

Waste is one of the most important sources of soil contamination. Waste can penetrate into the ground and pollute the water resources as well. Methods of waste disposal includes: dumping, incinerating and recycling.

In dumping method, areas are created as "land fill" and garbage or wastes is dumped there. In this method, waste is dumped below the ground level, aimed at not to be observed from ground surface but the said method creates subsequent problems. These problems include: pollution of water resources, producing bad odor and poisonous methane gas which provides fire danger, accumulation of harmful insects & organisms etc.

To control soil pollution caused by the waste, the following techniques are recommended:

- Application of effective technology for dumping waste like compressing and covering of openings & holes.
- Dumping waste higher than the highest underground water levels.
- Creating impenetrable layers in building of land fields.
- Creating drainage system for the collection of leachates.

- Using the gases produced in landfills.

In incineration method, all wastes are collected at a place away from residing place and then, they are put on fire. Incineration method is one of the worst methods of waste disposal, because, incineration will produce very poisonous gases which will pollute the air and incur irreparable loss to the environment as well. After incineration, ash of waste remains and will create visual pollution.

Recycling is the best method of waste disposal. With storing some waste and reusing them, human can greatly contribute to reduction of waste. In this method, not only creation of more waste is prevented, but cost will also be saved remarkably.

(c) Controlling Pollution Caused by Industrial Activities in Soil

This method includes all the pollutants which enter the soil by the discharges from factories. These waste are as follow:

Wastes produced by power plants, wastes of chemical industries, wastes of steel manufacturing industries, wastes of metal working industries, wastes of oil industries (extraction & refining), wastes of wood, cellulose & paper manufacturing industries, wastes of leather production industries as well as waste of food industries.

Accumulation of heavy metals in soil is the major concern of the industrial pollution. These metals include lead, cadmium, silver & mercury which have harmful effects on the organisms and have repeatedly caused environmental disease.

Some of these effects are as follows:

- Disturbance of biological activities of soil,
- Toxic effects on plants,
- Harmful effects on human being as a result of entrance of materials to the food chain.

Three main methods for soil decontamination from industrial wastes are as follows:

- Soil can be excavated up to the specified depth and the excavated soil can be taken away from the region & then, it can be restored.
- The soil can be restored at the same area.
- Keeping soil in the area is the other method. Under such circumstance, auxiliaries are added to the soil to prevent spread of infection to the plants, animals and human.

Usually, a large plastic is drawn on the soil to prevent spread of soil pollution, to prevent water from penetrating into the soil and to prevent spread of pollution of the other regions.

The soil restoration methods used are as follows:

Using water to remove pollutants from the soil, using chemical & aerial solvents, eliminating pollutants with incineration, helping natural organisms for breaking down atoms of pollutants, adding materials to the soil for preventing spread of pollution to the other regions.

Role of Plants in Controlling and Decreasing Soil Pollution (Phytoremediation)

Pollution caused by the exhaust of cars & homes and heating devices & appliances will incur irreparable damages to the health of human beings, animals, water, soil and air.

Increase of disease in human beings, eradication of many plants and endangerment of many generations of animals are a solid evidence of the said claim.

For this purpose, biological scientists and environmentalists are forced to think of natural ways (biological) for fighting with pollution. The use of plants is the simplest biological way, because, using plants is a non-technical and cost-effective method in terms of technology.

(d) Soil Pollution Control Technique Caused by Lead Existing in It

Fungi are used to fight lead existing in soil, because, satisfactory & good results have been received both in coexistence among plants and fungi like ARBASCULAR-MYCORRIZA FUNGI (AMF) in absorption of the lead. Creating colony of this fungus on the root of plants will cause increase of root level for absorption of the lead which results in more absorption of the lead element to the plant. These fungi help plants survive and tolerate pollution better.

The protective effects on fungi for the plants is as follow: MYCORIZA colony increased on root and all surface of the elements, the fungus play an intermediary role for absorption of heavy metals again. In lead mines, a type of fungus named Spurs Plows Niger can also be used. As compared with the other fungi, this fungus can absorb more lead from the environment without experiencing any damage in this respect. So, these types of fungi can help improve soil infected by the heavy metals like lead.

(e) Phytoremediation Technique to Control Soil Pollution

Phytoremediation is a cost-effective, environmental and scientific technique which is suitable for developing countries. Through the use of Green Plants Engineering like herbaceous and woody species, phytoremediation is used for removing pollutants from water, soil or decreasing risks of environmental pollution like heavy metals, rare elements, organic compounds & radioactive materials.

Heavy metals are the most important mineral pollutions and soil microorganisms are able to decompose organic pollutants. But for microbial decomposition of metals, there is a need for organic or metal changes, in which, plants are presently used for this part.

In phytoremediation method, plants are classified based on absorption mechanism and soil pollution by heavy metals is reduced through chemical, physical & biological methods. Removal of soil pollution is usually carried out by two methods: 1- inside the site and 2-outside the site. In outside-of-the-site methods, the contaminated soil is transferred to the other place and after removal of its pollution, it is turned to the first place. In another methods i.e. inside-the-site, there is no need to transfer or move the soil, rather, when pollutions are turned into organic, capability of their biological uptake is reduced. To reduce pollution of mineral pollutants in soil, the methods of organic, complex and soil pH increase by the lime can be used.

But most of these methods are expensive and will destruct the environment. In technology of using plants as phytoremediation, green plants and their relationship with soil microorganisms is used to reduce soil & groundwater pollution. This technology can be used for removing each of two pollutants of soil i.e. organic and inorganic.

The studies show that application of physico-chemical techniques will cause eradication of soil useful microorganisms like microriza nitrogen stabilizers and consequently, it will weaken biological activities of the soil which will cost dearly in comparison with the phytoremediation technique.

Soil and water plants are used in rizo-filtration methods in that pollutants of contaminated water resources are deposited or condensed with little density in their roots. This method is applied especially for industrial wastewater treatment plants, agricultural runoff and-or wastewater of acid mines and is suitable for the metals like lead, cadmium, copper, nickel, zinc & chromium. The plants like Indian mustard, sunflower, tobacco, rye and corn enjoy this capability. These plants enjoy high capability of absorbing lead from the sewage system, based on which, sunflower has the highest capability of absorbing lead from the sewage than other plants.

In another method, restricting mobility and availability of pollutants is carried out in soil through using power of the roots. This method is usually used in soil, sediment & sludge to reduce the pollution and is carried out through absorption, deposition, complex and/or reduction of the capacity.

In plant evaporation methods, plant absorb pollutants from the soil & then convert them into the steam and transfer with the transpiration & atmospheric operation. This method is used in growing trees to absorb organic and inorganic pollutants. In another method which is known as "Reduced Plan" the plant helps removal of pollution from soil & underground waters with its metabolism through transferring, decomposition, stabilization and sublimation of the pollutant compounds. In this method, organic compounds are broken into simpler molecules and can enter inside the plant tissue. The studies have shown that plants enjoy the enzymes that can decompose wastes of chlorinated solvents like tri-chlorine, ethylene and other insecticides.

Some heavy metals require little amount of living organisms although excessive increase of the same essential heavy metals can be harmful for the organisms. The unnecessary heavy metals include arsenic, antimony, cadmium, chrome, mercury and lead. These metals are very important with relation to the pollution of soil & surface waters and are taken into consideration in phytoremediation science.

Plants have three basic strategies for the growth in soils contaminated with the heavy metals.

1. The species which prevent entrance of metals to their aerial part or hold concentration of metals low in the soil.
2. The species which accumulate metals in their aerial organs and return them to the soil.
3. The plants which can condense metals in their aerial organs and also the plants which absorb high concentration of pollutants and condense in their root, stem or leaf.

Removal of Soil Oil Pollutions through Using Phytoremediation Technique

Oil Pollution is an inevitable consequence of rapid population growth and industrialization process. In the same direction, soil pollution by the oil hydrocarbon materials can be observed vastly around oil exploration and refining installations typically via transfer routes of these materials across oilfields.

In addition to the direct emission of these pollutant, the dust from burning gases with oil adds toxic and harmful materials to the soils of the region which have vast impact on ecosystem of the region.

Phytoremediation is one of the methods, which take advantage of plants and microorganisms along with them for cleaning up the contaminated environment (water & soil).

In fact, phytoremediation will provide suitable conditions for the growth and development of the plant as well as increase of natural cleanup activities through the application of human intervention such as agricultural technology (ploughing and fertilizers). Cost-effectiveness, inexpensiveness and its simplicity is the most important advantage of this method than any other methods.

Phytoremediation has vast application on organic, inorganic and radioactive materials. This process is sustainable, affordable and is suitable for the developing countries. Generally, this method is cost effective as well. The efficiency of this method is increased with the application of fast-grown plants with high absorption power of the heavy metals. In most contaminated places, appropriate species have been identified for the removal of pollution. Two methods of composting and condensing can be regarded as preliminary stages for reducing production volume of these plants, but it should be considered that leachate of condensation should be collected completely.

6 CHAPTER Short Questions & Answers on Soil Pollution

Q. What is soil and its significance?

Ans. The top layer of earth is called soil. More than 80% of the land is covered by soil and it is 3 to 4 meter in depth on the earth surface. It is one of the most important natural resources. It consists of a mixture of minerals, organic material, living organism, air and water that together support the growth of plant life. The traditional meaning of soil is the natural medium for the growth of land plants. Many terms are used to refer to the non-water surface of the earth. The soil scientist calls it soil, a geologist may call fragmental rock or overburden, an engineer may call land. Land is a general term to describe the non-water portion of the earth, usually in reference to a large surface area.

Significance of Soil

- The soil is the natural medium in which plants grow. It provides oxygen, water and minerals to plant roots.
- It supports life and forms the basis of plant's & animal's life.
- We get food, clothing and building material from the plants & crops that grow in soil.
- Soil is a habitat of numerous types of organisms, plants & animals and home for a number of friendly micro-organisms that maintain & enhance the fertility of the soil.

Q. Describe the Mechanism of the Formation of Soil.

Ans. Rocks of crust are continuously subjected to mechanical, chemical and biological actions. As a result, they are broken down into finer particles during long period of time. The process of breaking up of rocks into finer particles is called as weathering. Changes in temperature, beating effects of hails, rain water, pressure of winds & streams cause mechanical weathering. Addition or removal of oxygen within rock materials and action of carbonic acids cause chemical weathering. Sometimes, plants like lichens growing in rock crevices produce some chemicals. Those chemicals help in the breaking up of rocks into finer particles. Such type of weathering is called as biological weathering. The process of soil formation is called as Pedogenesis and study of the soil is called Pedology.

The weathered out fine particles get mixed with dead remains of plant & animals and with their excretory materials to form soil. Several factors contribute to the formation of soil from the parent material. This includes the mechanical weathering of rocks due to temperature changes & abrasion, wind, moving water, glaciers, chemical weathering activities, and lichens. Climate and time are also important in the development of soils. In extremely dry or cold climates soils develop very slowly, while in humid and warm climates soils develop more rapidly. Under ideal climatic conditions, soft parent material may develop into 1 cm of soil within 15 days. Under poor climatic conditions, a hard parent material may require hundreds of years to develop into soil. Fine particles(clays) in the surface soil layers are carried downward by water and lodge or deposit in layers at depth of from 15 to over 60 cm deep. Organic material accumulates mostly in upper surface layers of soils.

Q. What do you mean by soil horizon?

Ans. The various, distinguishable layers of soil that occur are called soil horizons.

Soil Horizon- A layer of soil or soil material approximately parallel to the land surface and differing from adjacent genetically related layers in physical, chemical, and biological properties or characteristics, such as color, structure, texture, consistency, amount of organic matter, & degree of acidity or alkalinity. They also tell much about the characteristics of a soil which include information about hardpans, depth of organic matter accumulation, soil denseness from clay deposition, and the extent of leaching.

A cross- sectional view of the horizons in a soil is called a 'soil profile'.The top layer or the surface litter layer, called the 'O-horizon', consists mostly of freshly-fallen & partially-decomposed leaves, twigs, animal wastes, fungi and other organic materials. Normally, it is brown or black.

The uppermost layer of the soil, called the 'A-horizon', consists of partially-decomposed organic matter (humus) and some inorganic mineral particles. It is usually darker and looser than the deeper layers. The roots of most plants are found in these two upper layers. As long as these layers are anchored by vegetation, the soil stores water and releases it in a trickle throughout the year instead of in a force like a flood. These two top layers also contain a large amount of bacteria, fungi, earthworms, and other small insects, which form complex food webs in the soil, help recycle soil nutrients, & contribute to soil fertility.

The 'B-horizon', often called the subsoil, contains less organic material and fewer organisms than the A horizon. The area below the subsoil is called the 'C-horizon' and consists of weathered parent material. This parent material does not contain any organic materials. The chemical composition of the C-horizon helps to determine the pH of the soil and also influences the soil's rate of water absorption & retention.

Q. Describe the classification of various types of soil.

Ans.

(A) Types of Soil on the Basis of Particle Size

On this basis of size of particles or according to their texture, soil is classified into three main types:

(a) sandy soil, (b) clayey soil, and (c) loamy soil.

Sandy Soils: A soil having loosely packed coarse or large sized particles is called as sand. Sandy soils are well aerated, warm and are easy to work or dig. This type of soil remains dry most of the time and their nutrients are quickly washed out. This is because, it cannot hold water.

Clayey Soil It made up of particles of very small sizes. Those particles remain compactly arranged. It is sticky and cohesive or waterlogged when wet. It can hold greater quantity of water.

Loamy soil: A soil which is made due to mixing of clay (20-30%), silt (30-45%) & sand, which make them fertile and easy to work, is called as loamy soil. A compact soil made up of rather small sized particles as compared to sand is called as silt. Loams are the best soils for plant growth

Sometimes, soils contain dead and decomposed parts of plants & animals. Such a decomposed organic matter found in the soil is called as HUMUS. It is crucial to maintaining the balance of minerals and nutrients needed for plant growth. Presence of humus in the soil is essential for high production of crops. Farmers add humus in the soil by mixing manure (compost, wormy compost or natural manure) to the soil.

(B) Types of Soil on the Basis of Origin

On the basis of origin, soils are classified into following major groups:

Residual Soil

Transported Soil

Mountainous Soil

Residual Soil

The soil which is formed due to weathering of rocks and is found adjacent to those rocks is called as residual soil. This soil is basically composed of mineral particles of different sizes. On the basis of presence of mineral components, the residual soil is further divided into: Red soil, Black soil and Laterite soil. Red soil is red in colour due to the presence of oxides of iron in it. Besides iron, it contains potassium in good quantities. The black soil is made up of ash and lava that come out during volcanic eruptions. This soil contains aluminium, iron, calcium & magnesium in good amounts but it lacks phosphorus & nitrogen. This soil is suitable for the cultivation of cotton crops. The laterite soil is formed due to weathering of open parts of rocks. It is red-yellow in colour. It contains hydrogenous oxides of iron and aluminium. This is usually acidic hence unfit for the growth of plants.

Transported Soil

The soil which is transported through gravity, flowing water, wind or glacier from one place and deposited on some other place is called as transported soil. This soil has further been put into four different categories on the basis of mode of its transportation. Those four different categories are- Colluvial, Alluvial, Sandy and Glacial or Drift soils.

Colluvial soil is formed due to deposit of soil in low areas due to landslides. Alluvial soil is formed due to deposits of particles contained in the water of fast water currents. This soil often contains different layers and this is principally composed of fine-rounded particles. The soil which is made of sand particles deposited by wind at a certain place is called as sandy or desert soil. Only desert plants(xerophytes) grow in this type of soil. Some soils are formed due to deposits of particles through glaciers. Those soils are called as glacial soils.

Mountainous Soil

The soil which is found on mountains, hill slopes, cliffs and valleys, is called as mountain soil. This type of soil is basically composed of pebbles, clay, shell and limestone. The lower part of this soil is usually calcareous in nature. This soil is found in north-east Himalayan regions of India.

Q. Define soil pollution. Why it is an issue of concern?

Ans. Soil pollution is defined as the build-up in soils of persistent toxic compounds, chemicals, salts, radioactive materials, or disease causing agents, which have adverse effects on plant growth and animal health.

Soil is the thin layer of organic and inorganic materials that covers the Earth's rocky surface.

The organic portion, which is derived from the decayed remains of plants and animals, is concentrated in the dark uppermost topsoil. The inorganic portion made up of rock fragments, was formed over thousands of years ago by physical and chemical weathering of bedrock. Productive soils are necessary for agriculture to supply the world with sufficient food.

Soil pollution is an issue of serious concern because it is the soil on which we live and get our food supply. Rapid industrialization, urbanization and consequent increase in population have resulted in soil pollution.

Soil pollution is quite different from air pollution and water pollution as in soil pollution, pollutants remain in direct contact with soil for a relatively longer period.

Soil pollution can also be defined as, "Undesirable changes in physical chemical and biological characteristics of soil, which are harmful for all living organisms".

Q. How do we categorize the sources of soil pollution?

Ans. The sources of land pollution can be categorized as under:

- Agriculture
- Quarrying and mining
- Sewage sludge
- Dredged spoils
- Household-domestic solid waste
- Demolitions and constructions
- Industrial activities.

Q. List out the Reasons of Soil Pollution.

Ans. Reasons are listed below:

- Increasing urbanization
- disposal of untreated wastes
- Indiscriminate use of Agrochemicals
- Unscientific mining
- Dumping Industrial Wastes
- Accidental Pollution/Leakages

- Outdated technology, inadequate treatment and safety management of chemicals & wastes
- Lack of Designed & Engineered Landfills

Q. What are the different ways by which the soil can become polluted.

Ans. Soil can get polluted by:

- Seepage from a landfill
- Discharge of industrial waste into the soil
- Percolation of contaminated water into the soil
- Rupture of underground storage tanks
- Excess application of pesticides, herbicides or fertilizer
- Solid waste seepage

Q. What are the Causes of Soil Pollution?

Ans. Soil pollution is caused by the presence of man-made chemicals or other alteration in the natural soil environment. This type of contamination typically arises from the rupture of underground storage tanks, application of pesticides, percolation of contaminated surface water to subsurface strata, oil & fuel dumping, leaching of wastes from landfills or direct discharge of industrial wastes to the soil. The most common chemicals involved are petroleum hydrocarbons, solvents, pesticides, lead and other heavy metals. The occurrence of this phenomenon is correlated with the degree of industrialization and intensities of chemical usage.

A soil pollutant is any factor which deteriorates the quality, texture and mineral content of the soil or which disturbs the biological balance of the organisms in the soil. Pollution in soil has adverse effect on plant growth.

Pollution in soil is associated with:

- Indiscriminate use of fertilizers
- Indiscriminate use of pesticides, insecticides and herbicides
- Dumping of large quantities of solid waste
- Deforestation and soil erosion

Q. What are the Natural causes of land pollution?

Ans. Land pollution occurs massively during earth quakes, landslides, hurricanes and floods. All cause hard to clean mess, which is expensive to clean, and may sometimes take years to restore the affected area. These kinds of natural disasters are not only a problem in that they cause pollution but also because they leave many victims homeless.

Table 6.1: Sources & Mechanisms of Land Pollution

S. N.	Sources	Reason/Cause/Mechanism..
1	Agriculture	(i) Accumulation of animal manure. (ii) Excessive input of chemical fertiliser and pesticides. (iii) Farm residues. (iv) Illicit dumping of tainted or harmful crops on land.
2	Quarrying and Mining	(i) Use of explosives to blow up mines (ii) Mining tailings and refuse. (iii) discharge of acidic effluents. (iv) Use of heavy machineries and large number of vehicles.
3	Sewage sludge	Untreated municipal sludges.
4	Dredged spoils	Improper method of dredging disposals.
5	House hold or Domestic solid waste.	(i) Improper and inefficient method of collection transportation & disposal of solid waste causes accumulation. (ii) Imporper sanitation system.
6	Demolition and Construction	(i) Non-biodegradable rubbles or debris create unsightly condition.
7	Industrial	(i) Discharge of untreated effluents on the land for disposal. (ii) Deposition of process and effluent treatment sludges. (iii) Release of gaseous pollutants such as SO_2 causing acid rains.

Q. What are common chemicals causing soil pollution?.

Ans. The most common chemicals involved in causing soil pollution are:

- Petroleum hydrocarbons
- Heavy metals
- Pesticides
- Solvents

Q. what are different Types of Soil Pollution?

Ans. Different types of soil pollution are:

- Agricultural Soil Pollution
  - (i) pollution of surface soil
  - (ii) pollution of underground soil

- Soil pollution by industrial effluents and solid wastes
  - (i) pollution of surface soil
  - (ii) disturbances in soil profile
- Pollution due to urban activities
  - (i) pollution of surface soil
  - (ii) pollution of underground soil

Q. How do biological agents, mining activities and radioactive pollutants cause soil pollution?

Ans. Biological Agents

The major sources of biological agents causing soil pollution are:

- Human excreta
- Animal and bird excreta
- Faulty sanitation
- Municipal wastes
- Biomedical wastes

Industrial parasites are among the most threatening biological agents.

Mining Activities

The top layer of soil is generally damaged or destroyed during mining which results in:

- Heavy metals and other toxic substances (may be found as impurities) can destroy beneficial microorganisms of the soil. There may be health hazard, since rate of retention and absorption of heavy metals is higher in infants & children.
- Loss of fertile land;
- Formation of huge mountains of wastes;
- Addition of harmful substances to the soil.

The uncontrolled mine fires may also destroy the productivity of the areas near mines.

Radioactive Pollutants

When rain containing radionuclide's falls on the soil, the radioactivity gets transferred to the soil due to its absorption by the soil particles. Rain also adds radioactive contaminants into water bodies thus affecting the aquatic flora and fauna.

Radioactive pollutants can cause a number of undesirable diseases of digestive system if they enter our body through food chain.

Huge amounts of radioactive substances result from following sources:

- Nuclear device explosions
- Nuclear testing laboratories
- Nuclear power plants and weapons testing

All these are responsible for radioactive fallout, further enhancing soil pollution. Nuclear pollution is a kind of physical pollution of the environment. Nuclear pollution differs from air, water & soil pollution in the respect that it can bring about physiological changes in the present and future generations of all living beings.

There is no safe dose of nuclear radiation because even the smallest dose of radiation has many deleterious effects. All the living organisms are directly or indirectly affected by a large variety of radiations.

Q. How does indiscriminate and overuse of fertilizers cause soil pollution?

Ans. Soil nutrients are important for plant growth and development. Plants obtain carbon, hydrogen & oxygen from air and water. But other necessary nutrients like nitrogen, phosphorus, potassium, calcium, magnesium, sulfur and more must be obtained from the soil. Farmers generally use fertilizers to correct soil deficiencies. Fertilizers contaminate the soil with impurities, which come from the raw materials used for their manufacture. Mixed fertilizers often contain ammonium nitrate (NH_4NO_3), phosphorus as P_2O_5, and potassium as K_2O. For instance, As, Pb and Cd present in traces in rock phosphate mineral get transferred to super phosphate fertilizer. Since the metals are not degradable, their accumulation in the soil above their toxic levels due to excessive use of phosphate fertilizers, becomes an indestructible poison for crops.

The over use of NPK fertilizers reduce quantity of vegetables and crops grown on soil over the years. It also reduces the protein content of wheat, maize, grams, etc., grown on that soil. The carbohydrate quality of such crops also gets degraded.

Q. How does indiscriminate and overuse of pesticides, insecticides and herbicides cause soil pollution?

Ans. Plants on which we depend for food are under attack from insects, fungi, bacteria, viruses, rodents & other animals, and must compete with weeds for nutrients. To kill unwanted populations living in or on their crops, farmers use pesticides. The first widespread insecticide use began at the end of World War II and included DDT (dichloro diphenyl trichloroethane) & gammaxene. Insects soon became resistant to DDT and as the chemical did not decompose readily, it persisted in the environment.

Since it was soluble in fat rather than water, it biomagnified up the food chain and disrupted calcium metabolism in birds, causing eggshells to be thin & fragile. As a result, large birds of prey such as the brown pelican, ospreys, falcons and eagles became endangered. DDT has been now banned in most western countries. Ironically many of them including USA, still produce DDT for export to other developing nations whose needs outweigh the problems caused by it. The most important pesticides are DDT, BHC, chlorinated hydrocarbons, organophosphates, aldrin, malathion, dieldrin, furodan, etc. The remnants of such pesticides used on pests may get adsorbed by the soil particles, which then contaminate root crops grown in that soil. The consumption of such crops causes the pesticides remnants to enter human biological systems, affecting them adversely.

An infamous herbicide used as a defoliant in the Vietnam War called Agent Orange (dioxin), was eventually banned. Soldiers' cancer cases, skin conditions and infertility have been linked to exposure to Agent Orange. Pesticides not only bring toxic effect on human and animals but also decrease the fertility of the soil. Some of the pesticides are quite stable and their bio- degradation may take weeks & even months.

Pesticide problems such as resistance, resurgence, and health effects have caused scientists to seek alternatives. Pheromones and hormones to attract or repel insects and using natural enemies or sterilization by radiation have been suggested.

Q. How does Dumping of solid wastes cause soil pollution?

Ans. In general, solid waste includes garbage, domestic refuse and discarded solid materials such as those from commercial, industrial & agricultural operations. They contain increasing amounts of paper, cardboards, plastics, glass, old construction material, packaging material and toxic or otherwise hazardous substances. Since a significant amount of urban solid waste tends to be paper and food waste, the majority is recyclable or biodegradable in landfills. Similarly, most agricultural waste is recycled and mining waste is left on site.

The portion of solid waste that is hazardous such as oils, battery metals, heavy metals from smelting industries and organic solvents are the ones we have to pay particular attention to. These can in the long run, get deposited to the soils of the surrounding area and pollute them by altering their chemical & biological properties. They also contaminate drinking water aquifer sources. More than 90% of hazardous waste is produced by chemical, petroleum & metal related industries and small businesses such as dry cleaners & gas stations contribute as well.

Solid Waste disposal was brought to the forefront of public attention by the notorious Love Canal case in USA in 1978. Toxic chemicals leached from oozing storage drums into the soil underneath homes, causing an unusually large number of birth defects, cancers and respiratory, nervous & kidney diseases.

Q. How does Deforestation causes soil pollution?

Ans. Soil Erosion occurs when the weathered soil particles are dislodged and carried away by wind or water. Deforestation, agricultural development, temperature extremes, precipitation including acid rain, and human activities contribute to this erosion. Humans speed up this process by construction, mining, cutting of timber, over cropping and overgrazing. It results in floods and cause soil erosion.

Forests and grasslands are an excellent binding material that keeps the soil intact & healthy. They support many habitats and ecosystems, which provide innumerable feeding pathways or food chains to all species. Their loss would threaten food chains and the survival of many species. During the past few years quite a lot of vast green land has been converted into deserts. The precious rain forest habitats of South America, tropical Asia and Africa are coming under pressure of population growth and development (especially timber, construction & agriculture). Deforestation is slowly destroying the most productive flora and fauna areas in the world, which also form vast tracts of a very valuable sink for CO2.

Q. How Soil Pollution is caused Due to Urbanization?

Ans. Pollution of surface soils:

Urban activities generate large quantities of city wastes including several Biodegradable materials (like vegetables, animal wastes, papers, wooden pieces, carcasses, plant twigs, leaves, cloth wastes as well as sweepings) and many non-biodegradable materials (such as plastic bags, plastic bottles, plastic wastes, glass bottles, glass pieces, stone / cement pieces). On a rough estimate Indian cities are producing solid city wastes to the tune of60,000 - 80,000 metric tons every day. If left uncollected and decomposed, they are a cause of several problems such as:

- **Clogging of drains:** Causing serious drainage problems including the burst / leakage of drainage lines leading to health problems.
- **Barrier to movement of water:** Solid wastes have seriously damaged the normal movement of water thus creating problem of inundation, damage to foundation of buildings as well as public health hazards.
- **Foul smell:** Generated by dumping the wastes at a place.
- **Increased microbial activities:** Microbial decomposition of organic wastes generate large quantities of methane besides many chemicals to pollute the soil and water flowing on its surface.
- When such solid wastes are hospital wastes they create many health problems, as they may have dangerous pathogen within them besides dangerous medicines, and injections.

Pollution of Underground Soil

Underground soil in cities is likely to be polluted by:

- Chemicals released by industrial wastes.
- Decomposed and partially decomposed materials of sanitary wastes.

Many dangerous chemicals like cadmium, chromium, lead, arsenic, selenium products are likely to be deposited in underground soil. Similarly underground soil polluted by sanitary wastes generate many harmful chemicals. These can damage the normal activities and ecological balance in the underground soil

Brief Causes of soil pollution due to urbanization is given below:

- Polluted water discharged from factories.
- Runoff from pollutants (paint, chemicals, rotting organic material) leaching out of landfill.
- Oil and petroleum leaks from vehicles washed off the road by the rain into the surrounding habitat.
- Chemical fertilizer runoff from farms and crops.
- Acid rain (fumes from factories mixing with rain).
- Sewage discharged into rivers instead of being treated properly.
- Over application of pesticides and fertilizers.
- Purposeful injection into groundwater as a disposal method.
- Interconnections between aquifers during drilling (poor technique).
- Septic tank seepage.
- Lagoon seepage.
- Sanitary/hazardous landfill seepage.
- Cemeteries.

- Scrap yards (waste oil and chemical drainage).
- Leaks from sanitary sewers.

Q. What are the main causes of Land Degradation?

Ans. Land Degradation is the temporary or permanent lowering of the productive capacity of the land.

Various reasons of land degradation are listed below:

- **Soil erosion**

  Soil erosion means wearing away of soil. It refers to loss of the superficial layer of the soil.

  The top layer of the soil is very fertile. Therefore, its removal degrades the quality of land making it less productive.

  Soil erosion is caused due to heavy rainfall, lack of vegetation cover, and soil mismanagement like, uncontrolled grazing by cattle, faulty method of surface drainage, removal of forest cover.

- **Salination**

  Salination refers to increase in the concentration of soluble salts in the soil. Poor irrigation practices and flood water results in accumulation of dissolved salts on the soil surface. In dry areas water evaporate quickly leaving behind a white crust of salts in soil on the surface. The higher concentration of salts severely affects the water absorption process of the plants, resulting into poor production.

- **Water Logging**

  Water logging may be due to surface flooding or due to high water table. Excessive irrigation practices may cause water logging due to rise in water table of the area. The productivity of water logged soil is severely affected due to lesser availability of plants.

- **Shifting Cultivation**

  Among many tribal communities a very peculiar practice of cutting down a selected portion of the forest and burning exist. Crops were cultivated for a few years till the fertility of soil is reduced. When the fields become sterile they were abandoned & a fresh patch of forest were burnt and fresh fields were formed for crop cultivation.

- **Urbanization**

  Man is to blamed for most of the land degradation. Productive areas are fast reducing because of developmental activities such as human settlement, industries, roads, railways, airports etc.

- **Dumping**

  Dumping of solid wastes is one of the most important factors which are responsible for the land pollution. In developing countries like India, the garbage and refuse products are simply dumped leading to land pollution.

Q. What are the Agricultural, Industrial and Urban effects of Soil Pollution?

Ans. Agricultural effects:

- Reduced soil fertility

- Reduced nitrogen fixation
- Increased erodibility
- Larger loss of soil and nutrients
- Deposition of silt in tanks and reservoirs
- Reduced crop yield
- Imbalance in soil fauna and flora

Industrial effects

- Dangerous chemicals entering underground water
- Ecological imbalance
- Release of pollutant gases
- Release of radioactive rays causing health problems
- Increased salinity
- Reduced vegetation

Urban effects

- Clogging of drains
- Inundation of areas
- Public health problems
- Pollution of drinking water sources
- Foul smell and release of gases
- Waste management problems

Q. What are the Environmental Long Term Effects of Soil Pollution?

Ans. When it comes to the environment itself, the toll of contaminated soil is even more dire. Soil that has been contaminated should no longer be used to grow food, because the chemicals can leach into the food and harm people who eat it. If contaminated soil is used to grow food, the land will usually produce lower yields than it would, if it were not contaminated. This, in turn, can cause even more harm because a lack of plants on the soil will cause more erosion, spreading the contaminants onto land that might not have been tainted before.

In addition, the pollutants will change the makeup of the soil and the types of microorganisms that will live in it. If certain organisms die off in the area, the larger predator animals will also have to move away or die because they've lost their food supply. Thus it's possible for soil pollution to change the whole ecosystems.

Effects of soil pollution in brief

- pollution runs off into rivers and kills the fish, plants & other aquatic life.
- crops and fodder grown on polluted soil may pass the pollutants on to the consumers.
- polluted soil may no longer grow crops and fodder.
- Soil structure is damaged (clay ionic structure impaired).
- corrosion of foundations and pipelines.
- impairs soil stability.
- may release vapours & hydrocarbon into buildings and cellars.

- may create toxic dusts.
- may poison children playing in the area.

Q. What are the Health Consequences of soil Pollution?

Ans. Land Pollution affects humans and wild life in number of ways such as:

- It can cause Respiratory problems
- Lead to birth defects
- Cause skin problems
- Causes cancers
- The toxic substances causing soil/ land pollution can get into human body directly by skin contacts
- Toxic substances being washed into water sources like rivers, reservoirs, pollutes underground water courses
- Consumption of vegetables and fruits grown on polluted soils
- Inhalation of Polluted dust or particles

Health Effects of Land Pollution are enumerated below:

Cancer

Many of the widely used pesticides on agricultural lands are potentially carcinogenic.

Neurological Problems

The health effects of soil pollution, of course, vary with the contaminant. Some toxins, such as lead, are both naturally occurring and come from manmade sources. Children are most vulnerable to the health effects of lead in the soil. Their developing neurological systems put them at risk for serious health issues, causing possible brain and nervous system damage.

Tissue Damage and Irritation

Mercury levels in the soil can increase through deposition from precipitation. Mercury can cause irritation of internal & external tissues, such as skin and the lining of gastrointestinal tract. The severity depends upon the toxin and length of exposure.

Reduction of Soil Health

Soil pollution also degrades soil health, impacting both the environment as well as people in the long run. The same chemicals that can harm people may also affect plants and micro-organisms. Acid rain, for example, can create toxic conditions by mobilizing what might otherwise be harmless chemicals in the soil. These toxins in turn can slow plant growth as well as impact the soil bacteria and other organisms necessary for plant growth.

Q. Describe Various Causes And Effects Of Soil Pollution.

Ans. Various causes and effects of soil pollution are listed below:

Discharge of Industrial Wastes:

Most of the industries produce wastes of one kind or the other such as harmful gases and chemicals with polluted water discharged from factories. These wastes are being generated and dumped on the land, mainly include the following:

- Toxic pollutants
- Heavy metals
- Organic compounds
- Inorganic complexes
- Non-biodegradable materials

The industrial pollutants which pollute air and water also pollute the soil.

Urban Wastes

Urban activities generate large quantities of wastes including several Biodegradable materials (like vegetables, garbage, waste paper, animal wastes, papers, wooden pieces, street sweepings, plant twigs, leaves, cloth wastes as well as packing materials) and many non-biodegradable materials (such as plastic bags, plastic bottles, plastic wastes, glass bottles, glass pieces, metallic cans, stone / cement pieces). If left uncollected, they are a cause of several problems such as:

- Clogging of drains: Causing serious drainage problems leading to health problems.
- Barrier to movement of water: Solid wastes have seriously damaged the normal movement of water thus creating problem of inundation, damage to foundation of buildings as well as public health hazards.
- Foul smell: Generated by dumping the wastes at a place.
- Increased microbial activities: Microbial decomposition of organic wastes generate large quantities of methane ; besides many chemicals are also responsible to pollute the soil and water flowing on its surface.
- When such solid wastes are hospital wastes they create many health problems: As they may have dangerous pathogen within them besides dangerous medicines, injections etc. .

Underground soil in cities is likely to be polluted by:

- Chemicals released by industrial wastes: Many dangerous chemicals like cadmium, chromium, lead, arsenic, selenium products are likely to be deposited in underground soil. These can damage the normal activities and ecological balance in the underground soil.
- Decomposed and partially decomposed materials of sanitary wastes may pollute underground soil, since these wastes generate many harmful chemicals.

  Urban wastes are also dangerous like industrial wastes because some of these do not degrade easily. Over population and increasing consumption have totally changed the very complexion of urban wastes into a complex mixture of food remains, papers, paints, varnishes and many other toxic chemicals.

Improper Agricultural Practices:

Indiscriminate use of fertilizers:

Soil nutrients are important for plant growth and development. Plants obtain carbon, hydrogen & oxygen from air and water. But other necessary nutrients like nitrogen, phosphorus, potassium, calcium, magnesium, sulfur and more must be obtained from the soil. The three primary soil nutrients often in short supply are potassium, phosphorus and nitrogen compounds. These are called macronutrients.

Certain other elements like boron, zinc and manganese are necessary in extremely small amounts and are known as micronutrients. When crops are harvested, a large amount of macronutrients and a small amount of micronutrients are removed with the crops. If the same crop is grown again, depleted levels of these nutrients can result in decreased yields. These necessary nutrients can be returned to the soil through the application of fertilizers. Farmers generally use fertilizers to correct soil deficiencies. Fertilizers contaminate the soil with impurities, which come from the raw materials used for their manufacture.

Mixed fertilizers often contain ammonium nitrate (NH_4NO_3), phosphorus as P_2O_5, and potassium as K_2O. For instance, As, Pb and Cd present in traces in rock phosphate mineral get transferred to super phosphate fertilizer. Since the metals are not degradable, their accumulation in the soil above their toxic levels due to excessive use of phosphate fertilizers, becomes an indestructible poison for crops. The over use of NPK fertilizers reduce quantity of vegetables and crops grown on soil over the years. It also reduces the protein content of wheat, maize, grams, etc., grown on that soil. The carbohydrate quality of such crops also gets degraded. Excess potassium content in soil decreases vitamin C and carotene content in vegetables and fruits. The vegetables & fruits grown on over fertilized soil are more prone to attacks by insects and disease.

Indiscriminate use of pesticides, insecticides and herbicides

Pesticides are a general class of chemicals or biological agents used to control, repel, attract or kill pests. Pests are organisms that include insects, weeds or other plants, birds, mammals, fish and microorganisms that compete with humans for food, destroy property, spread disease or are considered a nuisance. They are categorized according to their target use. The three major groups of pesticides are herbicides (weed control), insecticides (insect control), and fungicides (disease control). Nematocides are pesticides used to control soil, leaf, and stem dwelling nematodes (round worms).

Pesticide can consist of synthetic, organic, inorganic compounds or biological organisms. Organic pesticides contain carbon, while inorganic pesticides contain carbon which can be found only under the form of carbonate or cyanide. These pesticides are derivatives made from arsenic, mercury, fluorine, sulphur and copper, as well as derivatives made from cyanide. Organic pesticides can be divided into three groups: synthetic pesticides (developed in laboratories and manufactured), natural pesticides (from animal, microbial or vegetal origin) and microorganisms. Inorganic pesticides are mostly derived from minerals.

Pesticides are not a modern invention. Elemental sulfur was used by ancient Sumerians to protect their crops from insects. Medieval farmers and scientists experimented with chemicals ranging from arsenic to lead on common crops. In 1938, Dichloro-Diphenyl- Trichloroethane, or DDT was discovered to be extremely effective and rapidly became the most widely used insecticide in the world. The most important pesticides are DDT (Dichloro-Diphenyl- Trichloroethane), BHC (Benzene Hexa Chloride), chlorinated hydrocarbons, organophosphates, aldrin, malathion, dieldrin, furodan, etc.

Plants on which we depend for food are under attack from insects, fungi, bacteria, viruses, rodents and other animals, and must compete with weeds for nutrients. To kill unwanted populations living in or on their crops, farmers use pesticides. The remnants of such pesticides used on pests may get adsorbed by the soil particles, which then contaminate root crops grown in that soil. The consumption of such crops causes the pesticides remnants to enter human biological systems, affecting them adversely. Pesticides not only bring toxic effect on human and animals but also decrease the fertility of the soil.

DDT, one of the first synthetic organic insecticides to be used, was thought to be the perfect insecticide. DDT has a half-life (the amount of time required for half of the chemical to decompose) of 10-15 years. This means that if 100 kg of DDT were to be sprayed over an area, 50 kg would still be present in the area 10-15 years later. The use of DDT has been banned in some countries. However, India still permits the use of DDT, although only for purposes of mosquito control. Persistent pesticides become attached to small soil particles which are easily moved by wind and water to different parts thus affecting soils elsewhere. Persistent pesticides may also accumulate in the bodies of animals, and over a period of time increase in concentration if the animal is unable to flush them out of its system, thus leading to the phenomenon called bioaccumulation. When an affected animal is eaten by another carnivore, these pesticides are further concentrated in the body of the carnivore. This phenomenon of acquiring increasing levels of a substance in the bodies of higher trophic level organisms is known as 'bio magnification'. This process, especially in the case of insecticides like DDT, has been proved to be disastrous. DDT is a well-known case of bio magnification in ecosystems. The short-term and the long-term health effects to the persons using the pesticide and the public that consumes the food grown by using the pesticides are also major concerns. Exposure to small quantities of pesticides over several years can cause mutations, produce cancers, etc.

DDT is still an excellent insecticide, but it has two major disadvantages: (1) Its half-life in the environment is to long (10-15 years), and (2) it is bio accumulated in the fat of animals, since it is soluble in fat rather than water. Serious concerns about the human safety and biological impacts of DDT has led 100 countries to ban its use including India.

We should replace chemical fertilizers and pesticides by developing an alternative. Alternative agriculture is the broadest term that includes all non-traditional agricultural methods and encompasses sustainable agriculture, organic agriculture, alternative uses of traditional crops, alternative methods for raising crops, etc. Sustainable agriculture advocates the use of methods to produce adequate safe food in an economically viable manner, while maintaining the state of the ecosystem. Organic agriculture advocates avoiding the use of chemical fertilizers and pesticides.

Acceptable pesticides must have at least the following characteristics:

- They must be short lived in the environment. They must not exist long enough to bioaccumulate and biomagnify in the food chains. The residues must be gone from foodstuffs. Usually, it must be dissipated within a week or two of application time. To allow workers into fields, it is desirable for many pesticides to be gone within a few days.
- Pesticides must not be carcinogenic, teratogenic, or mutagenic. But there is no easy nor clear cut way to determine these characteristics with certainty.

- The pesticide must be effective yet able to be safely handled. Capable of safe handling means that the chemical is retained on the sprayed or treated area, is not overly volatile so that is moves extensively into the air, and can be safely handled by the applicator when reasonable care is used.

Excess Salts and Water

Irrigated lands can produce higher crop yields than those that only use rainwater. However, this has its own set of ill effects. Irrigation water contains dissolved salts and in dry climates much of the water in the saline solution evaporates leaving its salts, such as sodium chloride in the topsoil. The accumulation of these salts is called salinization, which can stunt plant growth, lower yields and eventually kill the crop & render the land useless for agriculture. These salts can be flushed out of the soil by using more water. This practice, however, increases the cost of crop production and also wastes enormous amounts of water. Flushing out salts can also make the downstream irrigation water saltier.

Another problem with irrigation is waterlogging. This occurs when large amounts of water are used to leach the salts deeper into the soil. However, if the drainage is poor this water accumulates in underground and gradually raising the water table. The roots of the plants then get enveloped in this saline water and eventually die. Thus, in the long run it is better for us to adopt sustainable farming practices to prevent the degradation of soil.

Dumping of Solid Wastes

In general, solid waste includes garbage, domestic refuse & discarded solid materials such as those from commercial, industrial and agricultural operations. They contain increasing amounts of paper, cardboards, plastics, glass, old construction material, packaging material and toxic or otherwise hazardous substances. Since a significant amount of urban solid waste tends to be paper and food waste, the majority is recyclable or biodegradable in landfills. Similarly, most agricultural waste and mining waste is left on site. The portion of solid waste that is hazardous such as oils, battery metals, heavy metals from industries have needed to pay particular attention. These can in the long run, get deposited to the soils of the surrounding area and pollute them by altering their chemical & biological properties. They also contaminate drinking water aquifer sources.

Q. Describe briefly the overall effects of soil pollution.

Ans. The effects of pollution on soil are quite alarming and can cause huge disturbances in the ecological balance & health of living creatures on earth. Some of the soil pollution effects are described below:

- Excessive use of fertilizers does not allow microbial flora in soil to flourish;
- Excessive use of nitrogen and phosphorus fertilizers makes the soil deficient in other micro-nutrients such as zinc, cobalt, etc.;
- Soil fertility is reduced or adversely affected, if pesticides remain in soil for longer duration;
- Pesticides like DDT, dieldrin, etc. are known to seep gradually through soil into groundwater and thus contaminate public drinking water supplies;

- Decrease in soil fertility and therefore decrease in the soil yield;
- Loss of soil and natural nutrients present in it. Plants also would not thrive in such a soil, which would further result in soil erosion;
- Disturbance in the balance of flora and fauna residing in the soil;
- Increase in salinity of the soil, which therefore makes it unfit for vegetation, thus making it useless and barren;
- Generally crops cannot grow and flourish in a polluted soil. Yet if some crops manage to grow, they would be poisonous enough to cause serious health problems in people consuming them;
- Creation of toxic dust leading is another potential effect of soil pollution;
- Foul smell due to industrial chemicals and gases might result in headaches, fatigue, nausea, etc. in many people;
- Soil pollutants would bring in alteration in the soil structure, which would lead to death of many essential organisms in it. This would also affect the larger predators and compel them to move to other places, once they lose their food supply; and
- Soil pollution can lead to water pollution if toxic chemicals leach into groundwater, or if contaminated runoff reaches streams, lakes, or oceans. Soil also naturally contributes to air pollution by releasing volatile compounds into the atmosphere.

Q. How can we control soil pollution?

Ans. The following steps have been suggested to control soil pollution.:

- To help prevent soil erosion, we can limit construction in sensitive area. In general, we would need less fertilizer and fewer pesticides.
- If we could all adopt the three R's: Reduce, Reuse, and Recycle, this would give us less solid waste.
- Reducing chemical fertilizer and pesticide use. Applying bio-fertilizers and manures can reduce chemical fertilizer and pesticide use.
- Biological methods of pest control can also reduce the use of pesticides and thereby minimize soil pollution.
- Reusing of materials such as glass containers, plastic bags, paper, cloth etc. at domestic levels rather than being disposed, thereby reducing solid waste pollution.
- Recycling and recovery of materials-This is a reasonable solution for reducing soil pollution. Materials such as paper, some kinds of plastics and glass can and are being recycled. This decreases the volume of refuse and helps in the conservation of natural resources. For example, recovery of one tonne of paper can save 17 trees.
- **Reforesting:** Control of land loss and soil erosion can be attempted through restoring forest and grass cover to check wastelands, soil erosion and floods. Crop rotation or mixed cropping can improve the fertility of the land.
- **Solid waste treatment**: Proper methods should be adopted for management of solid waste disposal. Industrial wastes can be treated physically, chemically and biologically until they are less hazardous. Acidic and alkaline wastes should be

first neutralized; the insoluble material if biodegradable should be allowed to degrade under controlled conditions before being disposed.

As a last resort, new areas for storage of hazardous waste should be investigated such as deep well injection and more secure landfills. Burying the waste in locations situated away from residential areas is the simplest and most widely used technique of solid waste management. Environmental and aesthetic considerations must be taken into consideration before selecting the dumping sites.

Incineration of other wastes is expensive & leaves a huge residue and adds to air pollution. Pyrolysis is a process of combustion in absence of oxygen or the material burnt under controlled atmosphere of oxygen. It is an alternative to incineration. The gas and liquid thus obtained can be used as fuels. Pyrolysis of carbonaceous wastes like firewood, coconut, palm waste, corn combs, cashew shell, rice husk paddy straw and saw dust, yields charcoal along with products like tar, methyl alcohol, acetic acid, acetone & a fuel gas.

Anaerobic/aerobic decomposition of biodegradable municipal & domestic waste is also being done and gives organic manure. Cow dung which releases methane into the atmosphere, should be processed further in 'gobar gas plants' to produce 'gobar gas' and good manure.

Q. List out steps to Reduce Soil Pollution.

Ans. Following steps can be taken to stop / reduce /avoid land pollution:

- Adoption of organic farming instead of using synthetic chemicals.
- Adoption of recycle and reuse of a waste materials.
- Proper disposal of industrial wastes into secured landfill sites.
- Educating people not to litter.
- Proper collection, transportation and disposal of municipal wastes.

Q. Briefly discuss the facts related to soil pollution.

Ans. Soil pollution refers to an increased level of harmful human made chemicals in the natural soil environment.

Industrial pollutants and pesticides are two major soil pollution sources.

Soil pollution can lead to serious water contamination and therefore create huge health issues.

Cleaning soil pollution is expensive and (usually) long-term process.

Heavy agricultural machinery used by many farmers makes soil more susceptible to pollution because it negatively affects the density and compactness of soil.

If not treated properly soil pollution could significantly reduce crop yields in years to come which could lead to future food shortages..

Soil pollution is especially big issue in China where contaminated water is often being used to irrigate very large amount of arable land.

If soil is polluted with heavy metals such as chromium and lead this can lead to higher incidence of cancer and leukemia in people.

The excessive soil pollution can also cause major problem for ecosystems by affecting the lower pyramid levels of the food chain and disrupting its proper functioning.

There are several different options used to clean polluted soil. These include excavating soil and taking it to a disposal site, aeration of soils, thermal and bio remediation, etc.

Soil pollution can also lead to soil erosion by decreasing the level of natural nutrients in the soil. Such decrease would cause reduced plant life in given region and without enough plants this would further result in soil erosion.

Even the short term exposure to some harmful chemicals in soil can lead to headaches, nausea, fatigue and skin rashes.

Contaminated soil isn't a safe place to grow food on because the poisonous chemicals may reach into human body once people digest the affected food.

Preventing soil pollution is much easier than to clean it. This is the main reason why soil needs to be constantly tested for dangerous chemicals. Only clean and healthy soil can produce healthy food.

Q. What are the types & causes of soil erosion?

Ans. The loss of nutrient rich top soil or disturbances to the soil structure by some natural processes or anthropogenic activities is known as soil erosion. Soil erosion is a part of the soil pollution. Soil erosion is creeping death of the soil. When the rate of removal of soil exceeds the rate of soil formation, the soil erosion occurs.

Soil Erosion & Deforestation

Soil erosion can be defined as the movement of surface litter and topsoil from one place to another. While erosion is a natural process, often caused by wind and flowing water, it is greatly accelerated by human activities such as farming, construction, overgrazing by livestock, burning of grass cover, and deforestation.

The loss of the topsoil makes a soil less fertile and reduces its water holding capacity. The topsoil, which is washed away, also contributes to water pollution by clogging lakes and increasing the turbidity of the water, ultimately leading to the loss of aquatic life. For one inch of topsoil to be formed it normally requires 200-1000 years, depending upon the climate and soil type. Thus, if the topsoil erodes faster than it is formed, the soil becomes a non-renewable resource. Therefore, it is essential that proper soil conservation measures are, used to minimize the loss of the topsoil.

Forests and grasslands are an excellent binding material that keeps the soil intact and healthy. They support many habitats and ecosystems, which provide innumerable feeding pathways or food chains to all species. Their loss would threaten food chains and the survival of many species. During the past few years quite a lot of vast green land has been converted into deserts. Many scientists believe that a wealth of medicinal substances including a cure for cancer and aids, lie in the forest. Deforestation is slowly destroying the most productive flora and fauna areas in the world, which also form vast tracts of a valuable sink for CO_2.

The soil erosions are of two types:

(i) Natural or geological or background soil erosion

(ii) Accelerated soil erosion

The natural soil erosion takes place by normal or natural conditions without any interference of man. But the accelerated soil erosion is due to active interference of man & animals in disturbing the equilibrium in between the rate of soil formation and rate of soil loss.

The magnitude of soil erosion is governed by the factors like corrosivity of the eroding agent, erodibility of the soil, scope, steepness length, land use practices and conservation strategies employed.

Causes of soil erosion:

The process of soil erosions is caused by the following agents:

(a) Water erosion:

Water is powerful agent causing soil erosion either in the form of rain drops or in the form of surface flow. When the soil is lost like a thin covering from a large surface area, the process is known as sheet erosion. When the sheet erosion takes place with full momentum, leading to cutting of soil surface as thin channels, it is known as rill erosion. Several rills from the steep slope converge forming wider channel known as gully erosion.

(b) Wind erosion:

In dry (arid) regions, the soil is mainly sandy and the vegetation is grossly inadequate. In such places, wind causes loss of soil.

(c) Land slide or slip erosion:

When the rocks at cliffs face heavy rains, their weights undergo sharp rise due to gravitational forces. Such rocks with high weights slip over lower rocks causing land slide and thereby slip erosion.

(d) Stream bank erosion:

During flood, the rivers splash their water against banks and cut through them taking away soil from the bank.

(e) Deforestation and overgrazing:

Loss of natural vegetation due to deforestation & overgrazing leads to dry and dusty soil which can be taken away easily by wind and water.

(f) Mining erosion:

A lot of soil is also eroded due to mining and industrial activities.

Since soil is very precious for the existence of living organisms, it is the need of the hour to prevent the loss of soil by adopting suitable mechanisms.

The process of soil loss can be checked by the following means:

(i) Protection of soil from the severity of rain drops through afforestation.
(ii) Preventing water from concentrating and passing through the slopes.
(iii) Slowing down the momentum of water flow adopting various scientific techniques.
(vi) Growing vegetation covers to minimize wind velocity.
(v) Growing soil binders like grasses to prevent direct contact in between soil and eroding agents.

The contribution of different global agent towards total soil degradation (erosion) are as follows:

- Overgrazing accounts for 34.5% of the degraded area
- Deforestation accounts for 29.5%
- Agricultural activities accounts for 28.1 %
- Over exploitation of soil accounts for 7.0%
- Bio-industrial activities accounts for 1.2%

7 CHAPTER Multiple Choice Questions & Answers

1. Terracing is an effective method of soil conservation:
   (a) Desert areas
   (b) Hilly areas
   (c) Plains
   (d) Riverine areas
2. Which is the main causative factor for desertification?
   (a) Irrigated agriculture
   (b) Overgrazing
   (c) Tourism
   (d) Developmental activities
3. Deforestation may reduce the chances of:
   (a) Rain fall
   (b) Frequent cyclones
   (c) Erosion of surface soil
   (d) Frequent landslides
4. Deforestation causes:
   (a) Soil erosion
   (b) Pollution
   (c) No floods
   (d) None of these
5. One of the following crops is the most effective in controlling soil erosion:
   (a) Cotton
   (b) Pigeon pea
   (c) Green gram
   (d) Maize
6. Sheet erosion is caused by:
   (a) Fast running rivers
   (b) Wind
   (c) Heavy rains
   (d) Glaciers
7. Soil conservation means:
   (a) Prevention of spread of desert
   (b) To check soil erosion by wind and rains
   (c) To check soil erosion by afforestation
   (d) All of these
8. Soil conservation can best achieved by having:
   (a) Wind screens
   (b) Good plant covers
   (c) Restricted human activity
   (d) Low rainfall
9. Main indicator of soil erosion is:
   (a) Deforestation
   (b) Afforestation
   (c) Ozone layer
   (d) Goats and grazing animals

Answer

1 **(b)**	2 **(b)**	3 **(a)**	4 **(a)**	5 **(b)**
6 **(c)**	7 **(d)**	8 **(b)**	9 **(a)**	

10. Dark coloured soils are considered most fertile as they contain:
    (a) Gravel and sand
    (b) Sand and calcium
    (c) Clay and humus
    (d) Silt and earthworms
11. Which of the following prevents afforestation in deserts:
    (a) Goat and grazing animals
    (b) Local resident
    (c) Unfertility of soil
    (d) Low rainfall
12. Eroded soils are:
    (a) Rich in plant nutrients
    (b) Unaltered in plant nutrients
    (c) Devoid of plant nutrients
    (d) All of these
13. Fifth of June is celebrated as:
    (a) World environment day
    (b) Darwin's birthday
    (c) World health and hygiene day
    (d) World population day
14. Mulching helps in:
    (a) Moisture conservation
    (b) Weed control
    (c) Improvement of soil structure
    (d) Increasing soil fertility
15. The ecological equilibrium in mined areas can be achieved by:
    (a) Preventing soil erosion
    (b) Developing a vegetation cover
    (c) Preventing overgrazing
    (d) Introducing agriculture
16. Agrostology is related with the study of:
    (a) Agricultural growth
    (b) Epiphytes
    (c) Grasses
    (d) Nematode disease
17. What is the major cause of diminishing wild life?
    (a) Cannibalism
    (b) Habitat destruction
    (c) Felling of trees
    (d) Paucity of drinking water
18. Rotation of crop is essential for:
    (a) Increasing fertility of soil
    (b) Increasing quality of protein
    (c) Getting different kinds of crops
    (d) Increasing quality of minerals
19. Soil erosion can be reduced by:
    (a) Reducing over grazing
    (b) Planting proper plants
    (c) Making dams
    (d) All of those
20. Soil erosion can be prevented by:
    (a) Deforestation
    (b) Heavy rains
    (c) Afforestation
    (d) Over grazing
21. Which of the following plants is used as windbreak?
    (a) Delonix
    (b) Mangifera
    (c) Eucalyptus
    (d) Prosopis
22. "Chipko Movement" is related to:
    (a) Plant conservation
    (b) Project tiger
    (c) Animal breeding
    (d) Operation flood

Answer

10 **(c)**	11 **(d)**	12 **(c)**	13 **(a)**	14 **(a)**
15 **(b)**	16 **(c)**	17 **(b)**	18 **(a)**	19 **(d)**
20 **(c)**	21 **(d)**	22 **(a)**		

23. In order to maintain proper ecological balance:
    (a) The existing forests should be cleared and new ones should be planted.
    (b) Some quick growing annuals should be planted if a tree must be cut for other uses.
    (c) Tree must be cut whenever necessary because the underground part performs the useful purpose.
    (d) A tree should be planted in place of one to be cut.
24. Side effect of heavy and constant grazing is that the:
    (a) Upper layers of soil are loosened and are eroded.
    (b) Lower layers are made more compact and root respiration is interfered with.
    (c) Lower compact layers make root penetration difficult.
    (d) All of the above.
25. Soil erosion is prevented by:
    (a) Afforestation
    (b) Deforestation
    (c) Denudation
    (d) Overgrazing
26. Constant grazing and browsing in an area may ultimately result in the formation of:
    (a) Dense forest
    (b) Grassland
    (c) Bushy vegetation
    (d) Desert
27. The possible beneficial effect of the grazing animals is:
    (a) Addition of their excreta to the soil
    (b) Eradication of weeds
    (c) Removal of wild plants
    (d) Removal of wild animals
28. When young plants are grazed too heavily they get killed because:
    (a) The roots are starved
    (b) Of the mechanical injury to roots
    (c) Most of the aerial parts are eaten away
    (d) All of the above
29. Soil erosion can be prevented by:
    (a) Allowing herbivorous animals to graze excessively.
    (b) Growing plants to form a soil cover.
    (c) Making the land sloppy.
    (d) Increasing fertility.
30. The grazing animals can change the type of vegetation by:
    (a) Bringing in plant pathogens
    (b) Selective grazing
    (c) Cross pollination
    (d) Bringing the seeds of other plants
31. The cutting of trees from the forests of hill near a catchment area:
    (a) Will have no effect on causing floods in plain.
    (b) May cause flood in plains in rainy season.
    (c) Will have no effect on climatic condition of that area.
    (d) Will benefit the mankind for more area for cultivation.

Answer

23 **(d)**	24 **(d)**	25 **(a)**	26 **(d)**	27 **(a)**
28 **(a)**	29 **(b)**	30 **(b)**	31 **(b)**	

32. Grasses possess great regenerative power because they are:
    (a) Dominant species
    (b) Rhizomatous
    (c) Constantly grazed
    (d) Fast growing
33. To prevent wind and water erosion, the crop is sometimes harvested in such a way that a basal stub of the plant is left behind. This phenomenon is called:
    (a) Ley farming
    (b) Mulching
    (c) Basin listing
    (d) Strip cropping
34. Soil erosion in plains consequent to rain fall is initiated by:
    (a) Rill erosion
    (b) Sheet erosion
    (c) Gully erosion
    (d) All of these
35. The term riparian erosion is applied to:
    (a) Gully erosion
    (b) Slip erosion
    (c) Wind erosion
    (d) Stream bank erosion
36. It occurs under normal conditions:
    (a) Geological erosion
    (b) Surface creep
    (c) Both (a) & (b)
    (d) Accelerated erosion
37. Erosion of very fine particles is seen on account of:
    (a) Saltation
    (b) Suspension
    (c) Surface creep
    (d) Shifting of dunes
38. If the top soil has been eroded, we retire a land and grow over it:
    (a) Lichens
    (b) Blue-green algae
    (c) Legumes
    (d) Grasses
39. If the same crop is repeatedly grown in a field:
    (a) Nitrogen starvation may result.
    (b) Specific mineral deficiency may arise.
    (c) Soil will become prone to diseases.
    (d) Water level in soil will recede.
40. Bunds, dams and drains are constructed to prevent:
    (a) Stream bank erosion
    (b) Sheet erosion
    (c) Slip erosion
    (d) Gully erosion
41. Overall mineral level of soil recedes due to:
    (a) Leaching only
    (b) Utilization by plants
    (c) Both (a) & (b)
    (d) Chemical reactions
42. Odum has described soil erosion as:
    (a) Soil pollution
    (b) Creeping death of soil
    (c) Loss of living stratum
    (d) Loss of feeding zone
43. A slopy land on account of continued rainfall initially undergoes:
    (a) Rill erosion
    (b) Gully erosion
    (c) Sheet erosion
    (d) All of these

Answer

32 **(c)**	33 **(b)**	34 **(b)**	35 **(d)**	36 **(a)**
37 **(a)**	38 **(d)**	39 **(b)**	40 **(d)**	41 **(c)**
42 **(a)**	43 **(a)**			

44. The amount of soil lost to sea each year due to rainfall is around:
    (a) 3,000 million tons
    (b) 3,000,000 million tons
    (c) 3,00,00,000 million tons
    (d) 3,00,000 billion tons
45. The chief cause of accelerated erosion is:
    (a) Excessive rains
    (b) Wind storms
    (c) Overfelling and overgrazing
    (d) All of these
46. The soil erosion on slopes proceeds in the following manner:
    (a) Gully? Rill? Ravine
    (b) Rill? Ravine? Gully
    (c) Rill? Gully? Ravine
    (d) Ravine? Rill? Gully
47. Wave erosion is more prevalent and well-marked:
    (a) Along rivers
    (b) At the sea shore
    (c) In water falls
    (d) In all these places
48. Soil conservation is the process where:
    (a) Sterile soil is converted into fertile soil
    (b) Soil is aerated
    (c) Soil erosion is allowed
    (d) Soil is protected against loss
49. The plants commonly sown for crop rotation are:
    (a) Cajanus and Dalbergia
    (b) Cajanus and Aeschynomene
    (c) Dalbergia and trigonella
    (d) Trigonella and Trifolium
50. In dry farming initially the fertility of the land can be increased by:
    (a) Agrostology
    (b) Basin listing
    (c) Growing grazing fields
    (d) Animal husbandry
51. If there is a little slope in the field, the soil erosion can be prevented by:
    (a) Contour terracing
    (b) Contour strip cropping
    (c) Contour farming
    (d) All of these
52. The fertility level of any soil is generally determined by:
    (a) Aeration and hydration
    (b) Humus and mineral contents
    (c) Humus, mineral and hydration level
    (d) All of these
53. The most common method used for cultivation on hills:
    (a) Levelling
    (b) Ridge terracing
    (c) Bench terracing
    (d) Channel terracing
54. The disadvantage of ploughing in arid and semi-arid zones is:
    (a) Loss of soil water
    (b) Soil particles become smaller
    (c) Soil organisms are exposed
    (d) All of these
55. Gullies are removed by:
    (a) Pan breaking
    (b) Levelling
    (c) Ridge terracing
    (d) All of these

Answer

44 **(b)**	45 **(c)**	46 **(c)**	47 **(b)**	48 **(b)**
49 **(d)**	50 **(d)**	51 **(b)**	52 **(d)**	53 **(c)**
54 **(a)**	55 **(b)**			

56. An important agrostological technique to check soil erosion in the initial stages is:
    (a) Ley farming
    (b) Land retirement
    (c) Basin listing
    (d) Contour farming
57. Which of the following plants are used as wind breaks?
    (a) Zizpyhus and Calotropis
    (b) Dalbergia and Delonix
    (c) Delonix and Albizia
    (d) Morinda and Anona
58. The most significant and initial requirement for dry farming is:
    (a) Shallow ploughing
    (b) Development of textures
    (c) Animal husbandry
    (d) Crop selection
59. The soil permeability can be decreased due to blockers called:
    (a) Pans
    (b) Peds
    (c) Clods
    (d) None of these
60. The removal of weeds by ploughing is called:
    (a) Mulching
    (b) Fallowing
    (c) Tilling
    (d) Contour farming
61. Contour binding is done to check:
    (a) Rill erosion
    (b) Gully erosion
    (c) Ravine formation
    (d) Sheet erosion
62. The soil binding is done in the best way by:
    (a) Algae
    (b) Mycorrhizae
    (c) Grasses
    (d) All of these
63. When trees and shrubs are planted in long rows along streams, they are described as:
    (a) Wind breaks
    (b) Soil binders
    (c) Shelter belts
    (d) Basin blisters
64. Digging pits on the slopes just for the accumulation of run off is called as:
    (a) Contour terracing
    (b) Contour trenching
    (c) Beach terracing
    (d) None
65. Terracing is done in:
    (a) Desert areas
    (b) Hilly areas
    (c) Dry areas
    (d) Plain areas
66. What cause of land pollution produces large holes in the ground and can cause erosion?
    (a) Garbage
    (b) Landfills
    (c) Mining
    (d) Farming
67. What cause of land pollution uses chemicals such as pesticides and herbicides?
    (a) Garbage
    (b) Landfills
    (c) Mining
    (d) Farming

Answer

56 **(a)**	57 **(a)**	58 **(d)**	59 **(a)**	60 **(c)**
61 **(d)**	62 **(c)**	63 **(c)**	64 **(b)**	65 **(b)**
66 **(c)**	67 **(d)**			

68. How can land pollution harm the environment?
   (a) It can destroy the habitats of animals.
   (b) It can allow for harmful chemicals to enter the soil.
   (c) It can release greenhouse gases.
   (d) All of the above.

69. What do we call large areas where garbage is placed on the land?
   (a) Landfills
   (b) Mines
   (c) Trash lands
   (d) Garbage lands

70. What do we call trash when it eventually decays and becomes part of the environment?
   (a) Recyclable
   (b) Biodegradable
   (c) Decayable
   (d) Destructible

71. What is it called when you collect organic waste and store it so it breaks down into fertilizer?
   (a) Recycling
   (b) Biodegrading
   (c) Composting
   (d) Trashing

72. Around how long does it take Styrofoam to decay and become part of the environment?
   (a) 1 year
   (b) 5 years
   (c) 20 years
   (d) It will never fully decay

73. The three primary soil macronutrients are
   (a) carbon, oxygen, and water.
   (b) copper, cadmium, and carbon.
   (c) potassium, phosphorus, and nitrogen.
   (d) boron, zinc, and manganese.

74. Which of the following is negative effects on the soil and water due to conventional, mechanized farming practices?
   (a) Reduction in soil organic matter.
   (b) Soil erosion.
   (c) Leaching of pesticides and fertilizers into the groundwater.
   (d) all of the above.

75. What is a a soil horizon?
   (a) A factor influencing how soil is formed
   (b) A layer of soil
   (c) An organism found within the soil
   (d) A technique used to map soils

76. Which three layers form the soil profile?
   (a) Air. water and soil
   (b) Minerals, organic matter and living organisms
   (c) Clay, silt and sand
   (d) The topsoil, subsoil and parent material

77. How does a 'sandy' soil feel like to touch?
   (a) Sticky
   (b) Gritty
   (c) Smooth
   (d) none of these

Answer

68 **(d)**	69 **(a)**	70 **(b)**	71 **(c)**	72 **(d)**
73 **(c)**	74 **(d)**	75 **(b)**	76 **(d)**	77 **(b)**

78. Why is organic matter (humus) an important part of soil?
   (a) It helps to improve water infiltration.
   (b) It can break down organic pollutants.
   (c) It converts nitrogen in the air into nitrates used by plants.
   (d) It is rich in nutrients, which is important for fertility.
79. Which of the following is NOT a common reason why soil maps are used?
   (a) To determine the land drainage capabilities of an area.
   (b) To determine the suitability of soils for particular crops.
   (c) To identify soils and their properties.
   (d) To record how soils are used by people.
80. Approximately how many micro-organisms can be found in a teaspoonful of soil?
   (a) 4 billion
   (b) 50 million
   (c) 500,000
   (d) 1000
81. Which of the following creatures will you not find in the soil?
   (a) Earthworm
   (b) Springtail
   (c) Mite
   (d) Lemur
82. Which of the following is not a threat commonly faced by soils?
   (a) Soil erosion
   (b) Percolation
   (c) Deforestation
   (d) Climate change
83. What is soil erosion?
   (a) It is the process by which soil is formed.
   (b) A harmful process that involves the removal and transport of soil by wind and water.
   (c) A natural method of filtering harmful pollutants.
   (d) A process often referred to as the 'greenhouse' effect.
84. What effect can soil have on health, if eaten or inhaled?
   (a) Nothing - it is perfectly safe.
   (b) It can be good for our health.
   (c) It can have serious health implications such as cancer.
   (d) none of these.
85. The soil component which comes from the weathering of rocks constitutes
   (a) the inorganic component of the soil.
   (b) the organic component of the soil.
   (c) not necessary for plant growth.
   (d) none of these.
86. Which of the following does not contribute to the chemical composition of soils?
   (a) The chemical nature of underlying rocks.
   (b) Volcanic activity.
   (c) Rainfall.
   (d) None of these.

Answer

78 **(d)**	79 **(d)**	80 **(a)**	81 **(d)**	82 **(b)**
83 **(b)**	84 **(c)**	85 **(a)**	86 **(d)**	

87. Constituents of soils which are easily leached by rain and taken up by plants are stored
    (a) in all parts of the soil.
    (b) on the exchange sites of the organic and clay components.
    (c) in the silicate fraction of the soil.
    (d) only in the organic fraction of the soil.

88. The particle size of which of the following the smallest?
    (a) sand
    (b) silt
    (c) clay
    (d) gravel

89. Soils containing a greater proportion of which of the following tends to drain quickly than the others?
    (a) sand
    (b) silt
    (c) clay
    (d) gravel

90. Living organisms in the soil
    (a) contribute to the organic component of the soil.
    (b) do not have any role in soil formation.
    (c) facilitate the disintegration process of rocks in soil formation.
    (d) both a & c.

91. The soil horizon which is rich in minerals is
    (a) A - horizon
    (b) B - horizon
    (c) C - horizon
    (d) O - layer

92. What is the first step in pedogenesis?
    (a) disintegration of parent materials
    (b) formation of the parent material
    (c) decay of plant materials
    (d) formation of the soil horizons

93. Regolith is
    (a) bedrock.
    (b) soil formed by the weathering of rocks.
    (c) a layer of rock debris formed by the gradual weathering of bedrock.
    (d) the organic soil component.

94. The layer of soil from which the eluviation of clay, iron, & aluminum oxides take place by leaching is
    (a) A-horizon
    (b) B-horizon
    (c) C-horizon
    (d) E-horizon

95. Unhealthy soil management methods
    (a) degrades soil quality
    (b) enhances erosion
    (c) causes soil pollution
    (d) all a, b & c

96. Most common toxic soil pollutants may arise from
    (a) leaching of wastes from landfills.
    (b) Application of animal dung to the soil.
    (c) acid rain.
    (d) decaying plants.

Answer

87 **(b)**	88 **(c)**	89 **(a)**	90 **(d)**	91 **(c)**
92 **(b)**	93 **(c)**	94 **(d)**	95 **(d)**	96 **(a)**

97. Which of the following human activities does not increase the level of heavy metals in soils?
   (a) Industrial and agricultural activities
   (b) Waste incineration
   (c) Combustion of fossil fuels
   (d) None of these

98. Organic pollutants enter the soil through
   (a) bedrock weathering
   (b) decomposition of plant and animal debris
   (c) atmospheric deposition
   (d) all of the above

99. Pollution of agricultural soils by heavy metals can
   (a) increase agricultural yield by providing plants sufficient amount of essential metals.
   (b) introduce toxic heavy metals into the food chain.
   (c) increase the number of soil microorganisms which can boost agricultural yield.
   (d) all a, b & c.

100. Pick out the correct statement about soil.
   (a) Most soils have pH value between 1 & 5.
   (b) Lime is used to treat soil, that is too alkaline.
   (c) Loam is a poor soil which drains poorly and is low in nutrients.
   (d) Organisms like worms help circulate air in soil.

101. Which of the following is used to treat soil which is too alkaline?
   (a) Sand
   (b) Lime
   (c) Peat
   (d) Marble

102. Which of the following factors does not influence a soil's abi1ity to hold water nutrients and air?
   (a) Soil texture & structure
   (b) Soil porosity
   (c) Soil pH value
   (d) Soil composition

103. The three primary soil macronutrients are
   (a) carbon, oxygen, and water.
   (b) copper, cadmium and carbon.
   (c) potassium, phosphorus, and nitrogen.
   (d) boron, zinc, and manganese.

104. Which of the following has negative effects on the soil & water due to conventional, mechanized farming practices?
   (a) Soil compaction and soil erosion
   (b) Reduction in soil organic matter
   (c) Leaching of pesticides and fertilizers into the groundwater
   (d) All of the above

105. What are the two greatest causes of land degradation?
   (a) Industry and mining.
   (b) Crop growing and wood collection.
   (c) Energy production and pollution.
   (d) Overgrazing and deforestation.

Answer

97 **(d)**	98 **(c)**	99 **(b)**	100 **(d)**	101 **(c)**
102 **(c)**	103 **(c)**	104 **(d)**	105 **(d)**	

106. What is monoculture?
   (a) One culture being dominant in society.
   (b) Crops being grown by individuals for their own consumption.
   (c) Cultivation of plants in a greenhouse.
   (d) The growing of one crop over a large area of land.

107. Which is the top horizon in a soil profile?
   (a) The A-horizon.
   (b) The C-horizon.
   (c) The B-horizon.
   (d) The O-horizon.

108. How much fertile land is being lost due to degradation each year?
   (a) Seven million hectares.
   (b) Five million hectares.
   (c) Four million hectares
   (d) Six million hectares.

109. Which of the following are both types of soil erosion?
   (a) Gully and cliff.
   (b) Sheet and rill.
   (c) Frost shattering and exfoliation.
   (d) Plucking and abrasion.

110. The soil is:
   (a) A sedimentary rock
   (b) A layer of earth crust
   (c) Natural body, located in land surface with properties and specific functions
   (d) None of the above

111. Soil pollution means:
   (a) Any action that causes disruption of normal functioning of soil as a medium for life
   (b) Ground ozone
   (c) Ground nitration
   (d) Ground sulphonation

112. Normal functioning of soil disturbance occurs through:
   (a) Deterioration
   (b) Physical degradation, chemical and biological of soil
   (c) Osmosis
   (d) Hydrolysis

113. Soil disturbance means:
   (a) Increase of soil fertility
   (b) Lower level of nitrogen in the soil
   (c) Increase of soil carbon levels
   (d) Action that adversely affects the fertility of soil

114. Soil has the following vital signs:
   (a) Providing food
   (b) Providing drinking water
   (c) Ensuring habitable surface
   (d) Ensuring beautification of nature

115. The greatest loss is due to:
   (a) Increasing water
   (b) Erosion
   (c) Plant osmosis
   (d) Volcanic eruptions

116. Harmful effects can cause:
   (a) Nitrogen reached in the soil
   (b) Sulfur
   (c) Heavy metals from various sources that reached the ground
   (d) Excess soil water

Answer

106 **(d)**	107 **(d)**	108 **(a)**	109 **(b)**	110 **(c)**
111 **(a)**	112 **(b)**	113 **(d)**	114 **(a)**	115 **(b)**
116 **(c)**				

117. Unintended consequences of the long use of pesticides:
   (a) Yellowing leaves
   (b) Serious damage to various forms of life
   (c) Premature to mature fruit
   (d) Rapid development of plant

118. Soil and soil microorganisms associated:
   (a) Plays role of a universal biological purifier and neutralizer
   (b) Plays the role of pollutant
   (c) Plays the role of decorator
   (d) Plays the role of balancer

119. The meaning of the term "nitrogen" is:
   (a) Lifeless
   (b) Has no significance
   (c) Rich in chemical elements
   (d) With life

Answer

117 **(b)**	118 **(a)**	119 **(a)**

8 CHAPTER Glossary of Terms Related to Soil Pollution

Acidification

Lowering pH value of a soil with low buffering capacity, typically forest soil, due to acid rain.

Acid rain

Fossil fuels can release chemicals such as sulphur when they are burnt (as petrol is in a car, or coal is in a power station). These chemicals can dissolve in atmospheric water and make rainfall unnaturally acidic. This can cause many environmental problems.

Aerated soil

An aerated soil is a soil with a good movement of air through the soil structure. The opposite is a wet water logged soil, where the soil pores are filled with water.

Aggregates

Aggregates in the soil are just soil lumps of a range of differing sizes. The word is also often used to describe gravel used to build roads.

Alluvium

Alluvium is a deposit made by a river or running water. This leads to an 'alluvial deposit' forming in the water. Alluvial soils are rich agricultural lands. Glaciers may also deposit sedimentary material.

Anaerobic soils

Anaerobic soils have very little oxygen present - for instance the wet, water logged, marshy soils in a bog. While anaerobic conditions are important for some processes, such as bacterial reduction of nitrate to nitrogen, these conditions can also produce hydrogen sulphide, methane and other undesirable substances.

Anion

A negatively charged ion. The most commonly found anions in soil waters include bicarbonates, sulphates, carbonates, chlorides and nitrates.

Arid conditions

Arid conditions are dry. Where rainfall is less than about 200 mm/year an area is 'semi-arid', rainfall below this leads to arid conditions such as deserts.

Autotrophic (bacteria)

Autotrophic bacteria can take carbon in from the atmosphere through the fixation, reduction and incorporation of carbon dioxide.

Available soil water

This is the amount of water held in the soil root zone, within which the 'easily available soil water' is the amount of soil water easily available for plants for evapotranspiration; which can be extracted by the plants. Soil also contains a smaller fraction of water which is 'held' too tightly by the soil particles for plants to use - this makes up the 'total available soil water'. Expressed in mm.

Bacteria

A bacteria is a simple microscopic single-celled organism. There are many forms of bacteria which thrive in soil. Bacteria play a crucial role in the nutrient cycles in soil, helping to break down organic matter.

Bearing capacity

This is effectively the weight a soil can withstand before severe damage occurs to the structure of the soil. Bearing capacity varies throughout the year, for instance a very heavy tractor that causes no damage on dry soils may cause a lot of damage to wetted soils.

Bio piles

Spreading of contaminated soils on a lined treatment bed, where air and/or nutrients dissolved in water are flushed through the soil pile to optimize biodegradation of contaminants

Bioremediation

A method using micro-organisms to treat contaminated soils to transform hazardous chemicals to less toxic or non-toxic compounds; bioremediation processes for excavated contaminated soils include bio piles, land farming, composting and use of slurry reactors

Bioventing

Ventilation of a contaminated soil by pulling a vacuum to erhance aerobic biodegradation of contaminants, or air sparging which volatilize contaminants in the saturated zone and enhance aerobic biodegradation

Biotope

A Biotope is a distinct geographical area having similar environmental conditions, such as soil, climate and relief.

Brownfield sites

Places with contaminated soils damaged by pollution due to e.g. industrial activities and the infrastructure it supports.

Buffering capacity

Buffering capacity is the ability of the soil to reduce high alkalinity or acidity levels coming perhaps from pollution (e.g. acid rain). Chalky or limestone soils for instance are very alkaline and can neutralize acids more effectively than acid, peat soils.

Bulk density

The dry mass of soil per unit bulk volume of soil. Expressed as kg/m^3 or g/cm^3.

Calcicolous

Calcicolous are plants able to grow in very alkaline conditions (high pH).

Cartography

The overall term used to describe the process of producing maps characterizing geographical variation in some property of soil. Traditional maps are made on paper. However, increasingly maps are held and manipulated in computer form.

Cation

A positively charged ion. The most commonly found cations in soil waters include sodium, potassium, calcium and magnesium.

Chemical and physical on-site treatment

Technologies to deal with contaminated soil on-site including heating, i.e. thermal process, soil washing or chemical extraction, soil flushing, solidification/stabilization, and chemical destruction

Clay

That mineral fraction of the soil with particles less than 0.002 mm diameter. Yes, this is what you make pots from - but clay soil is rarely that pure - it is often mixed with particles of different textures.

Compost

Compost is a mixture of decomposing vegetable matter and animal remains, as well as manure, etc., which breaks down into the 'organic layer'. Gardeners create compost heaps to hold grass clippings and vegetable peelings etc. which rot down to a rich organic material used for fertilizing and conditioning the soil.

Composting

Microbiological degradation of organic waste, often in special containers, used to treat municipal solid waste, wastewater treatment sludge, to leave behind a soil like residue, the compost to be used for soil improvement and fertilization

Coniferous

Coniferous (or evergreen) trees, such as Larch, are ones who retain their leaves all year long. Their leaves are designed to withstand adverse weather conditions such as cold or drought. The opposite is a 'deciduous' tree.

Contaminated soils

Soils damaged by pollution due to e.g. deposit of waste, industrial activities, military activities, and the infrastructure it supports, including metallurgical factories, petrochemical industries, factories for paints or pharmaceuticals, petrol stations, car washing stations and landfills.

Denitrification

Denitrification is the mechanism that allows the return of nitrogen to the atmosphere thus completing the cycle: atmosphere-fixation in the soil-mineralization and nitrification for use by plants-denitrification and return to the atmosphere. Denitrification is the reduction of nitrate to gaseous nitrogen, enabling it to pass from the soil to the atmosphere again.

Deposition

Deposition is the natural process of deposits being laid down which ultimately become soils. An example of a deposited soil is 'alluvium' which is sediment laid down in water. The opposite is 'erosion'.

Desertification

Desertification is the change in arid and semi-arid regions from formerly productive land to desert. This is often the cause of erosion, particularly by wind, and climate changes exacerbated by poor land management (such as deforestation).

Drainage

Reducing water content in over-moist soils and meadow soils to allow agriculture.

Effective rooting depth

The soil depth from which a fully grown plant extracts most of the water needed for transpiration.

Erosion

Erosion is the wearing away of land or soil through one or more processes. The main causes of erosion include the actions of water (rills, inter-rill, gully, snowmelt and river & lake bank erosion), wind (desiccation and wind-blow), translocation (tillage, land levelling, harvesting of root crops, trampling & burrowing animals) and geological (internal subterranean erosion by groundwater, coastal erosion & landslides). Erosion can also be increased by poor land management such as overgrazing, deforestation or inappropriate use of mechanization (e.g. ploughing down a hill slope).

Eutrophication

Eutrophication describes the process where a waterbody, such as a lake or a soil solution, becomes loaded with dissolved nutrients. This can be natural, but is often due to pollution. Algal blooms can remove oxygen in the water, harming fish life.

Evaporation

This is the rate of water loss from liquid to vapour (gaseous) state from an open water, wet soil or plant surface, usually expressed in mm day^{-1}.

Evapotranspiration

The process whereby water is passed from liquid to vapour (gaseous) state through transpiration from vegetation, and evaporation from soil & plant surfaces.

Fauna and Flora

Fauna are the organisms that live in and near the soil, from ants, snails and slugs to fungal mycelia, mycorrhizae and algae. Flora are the plants living in soil.

Fertiliser

Substances added to the soil (in granules or liquid form) containing one or more recognized plant nutrients, such as nitrogen, potassium and phosphorus (called an 'NPK' mixture).

Fertilisers, which can be either organic or inorganic in nature are really plant foods, and are added to the soil to improve the quality or quantity of plant growth.

Field capacity

Field capacity describes soil when it is completely wetted (excepting free drainage), and where there is plenty of water for plant roots. This occurs after ample irrigation or rainfall, when the rate of downward movement of water has substantially decreased, usually 1-3 days after rain or irrigation. It is expressed as a mass or volume fraction of soil water or a depth of water per metre of soil or mm m^{-1}.

Fixation

Nitrogen fixation is the conversion of atmospheric nitrogen into organic nitrogen compounds which are then available to green plants; a process that can be carried out only by certain strains of soil bacteria (e.g. Rhizobium, Clostridium and Azotobacter).

Fossil fuels

Fossil fuels are the petrified remains of ancient plant and animal matter. Organic matter was compressed & heated in the earth's crust over geological time to form oil, coal and natural gas.

Fungicide

A fungicide is a chemical 'pesticide' designed to eradicate fungi that can grow on and spoil plants such as crops & fruit.

Geology

Solid material from which most soil derives, characterized by the horizon symbol 'R' for rock. Soil may lie on top of the geology it came from - or may be not if ancient glaciers pushed the soil around like a primeval bulldozer. The lower horizon may also be denoted 'C' where there are soft deposits such as alluvium. Geology is also a scientific field concerning the study of rock.

Germination

Germination is the first stage in the development of a plant from its seed. Seeds lie 'dormant' in the soil waiting for the right conditions (related to sustained favourable temperature and moisture levels).

Glacial till

Glacial till is sediment ground up beneath a glacier and deposited as 'sub-glacial till', or glacial till deposited at the 'terminus' as a 'terminal moraine'. Till is a jumble of rocks and finely ground material ground up by the movement of the glacier over the rock. Till deposited in this way leaves an unstratified soil of varying composition.

Gley soils

Gley (or gleyed) soils are soils developed under conditions of poor drainage, resulting in reduction of iron & other elements and also in a typical grey/blue soil colouring. There are two main types of gley soil: surface water gleys where water saturating the soil comes from surface drainage; and-ground water gleys where saturation is due to fluctuating groundwater levels. A third type, unripened gley soil forms in brackish flooding conditions (tidal creeks).

Global warming

Global warming is the slow warming up of the Earth's atmosphere, caused by the 'Greenhouse Effect'. Rising temperatures lead to melting polar ice in turn leading to rising sea-levels. Global warming will have a big impact on the soil resource worldwide.

Ground water remediation

Activities to decontaminate a soil which pollutes ground water, through ground water pumping to remove the pollutant, the traditional method, which may require pumping water for years or decades; soil vapor extraction, which removes contaminants from the unsaturated soil zone by pulling a vacuum; and bioventing or air sparging to enhance aerobic biodegradation of contaminants.

Herbicide

A herbicide is a type of chemical applied to eradicate certain types of unwanted vegetation, such as weeds.

Horizon

One of the layers that form in the soil profile as a result of soil-forming processes such as weathering. A horizon is often highly visible and distinct. In garden soils and where there has been much soil movement these deeper layers may be lost or harder to observe - equally, where lots of organic matter and compost has been added, garden soils will have a strong organic rich horizon.

Humus

Organic matter, also called 'humus', forms from the decay of leaves, plants and other life.

Immobilisation

Immobilisation is the process by which available inorganic forms of nitrogen are returned to the organic form and in so doing become unavailable again. Immobilised nitrogen is reasonably secure in the soil in this form and can continue to be recycled within the soil from organic-mineral-organic states again via the mineralisation, nitrification immobilisation stages.

Incineration

Burning of waste in combustion facilities at temperature 800 - 1,400 °C or more, in which hazardous wastes are completely destroyed because of high temperature and long retention time, applied to both municipal & organic industrial hazardous waste

Industrial solid waste

Solid waste form industry in particular, mining & quarrying waste, energy generation waste and construction industry waste

Insecticide

A chemical 'pesticide' designed to eradicate insect pests on plants such as crops and fruit.

Infiltration

Infiltration is the movement of water from the surface down into the soil before moving down to the aquifers, or out to rivers. A portion of soil water may also be lost via the process of evapotranspiration.

Infiltration rate

The infiltration rate is the speed at which water can pass into the soil. You can imagine that with wet clay, this is slower than with dry sand.

Land reclamation

Improvement of land by protection & improvement of soil, e.g. by plantation to bind soil and reduce erosion, increase organic content in soil and balance water content.

Land farming

Spreading of contaminated soils on the surface of the ground to enhance natural microbial degradation of contaminants in the soil often successful in treating petroleum refinery waste in areas where air emissions do not pose unacceptable levels of health risks.

Landfill

An organised dumping of waste on land, as an effective means of solid waste disposal, both for municipal and industrial waste, where the waste is put on or in the ground and covered by earth.

Leaching

Leaching is the process where soluble materials (including nutrients and salts) in the soil are washed down the soil profile by water.

Lime and liming

Farmers add lime to improve their soils and to 'buffer' acidity. The production of lime involves limestone blocks roasting in lime kilns to produce the powdery white lime which is then spread on fields. Modern 'fertilisers' also contain lime.

Lithosphere

The Lithosphere is the outer rock shell of the planet, extending down several kilometres into the earth from the surface - the rocks and sediments well below the surface of the land & oceans.

Loam

Loam is a soil which contains clay, silt and sand as well as a mixture of organic material. Keen gardeners love loam - it's the best soil for potting plants and growing seeds in the greenhouse. Farmers like loam too as they are ideal for agricultural crops.

Macronutrients

These are nutrients which are needed in large quantities by plants - such as calcium, magnesium and potassium.

Manure

Manure is the well-rotted remains of dung, say from horses, which can be applied as a rich organic fertiliser to garden soil. Note, never add fresh dung directly to garden soils but only well-rotted remains

Marl

Marl is a calcareous clay, comprising about 30 to 65 % calcium carbonate ($CaCO_3$). Marl is most often found in old fresh water basins, swamps, or as the sediment of shallow lakes.

Micro fauna

Micro fauna is the smallest of the soil 'fauna' or animals, and are less than 0.1 mm in size, so you need a microscope to see them. The two most important ones are the nematodes and the protozoa.

Micro flora

Micro flora is tiny microscopic 'flora' in soils, such as bacteria, fungi and viruses. Microflora are extremely numerous in soils with billions in just one gram of soil and many thousands of species also within a single gram.

Micronutrients

These are nutrients which plants need only in small (or 'trace') amounts - such as iron, zinc, copper and boron.

Microorganism

Microorganisms are tiny organisms visible only with a microscope and include bacteria, protozoa, algae, & fungi. Note that viruses are not truly living organisms, but are sometimes classed as microorganisms. The soil is literally teeming with microorganisms.

Mineralisation

Mineralisation is the conversion of 'bound' organic nitrogen into the mineral (ionic or inorganic) form required for plant uptake. The inorganic nitrogen is then converted in a process known as nitrification from ammonium salts to nitrites, and from nitrites to nitrates by nitrifying bacteria which obtain their energy from this oxidation process. Once free nitrate is formed, the rapid recycling process offers many options, the nitrate may be assimilated by plants, denitrified or leached.

Mineral

An inorganic component of soil deriving from rock material, including substances such as calcium, iron, zinc, iodine and fluorine. There are also 'trace elements' such as selenium, manganese, molybdenum, chromium, cobalt, silicon, vanadium and nickel.

Mineral soils

Soils with less than 10% of organic matter, the most common soil type.

Mining and quarrying waste

Accounting for the largest amounts of solid waste, about 70% in the region, consisting of refuse from mining metal ores, hard coal, lignite, oil-shale and peat.

Moraine

A moraine is a form of glacial till, a soil sediment deposited at the terminus of a melting glacier. The moraine is a single, large mound deposited, typically at the edge of a glacier, being a jumble of rocks & finely ground material carried by the glacier and deposited when the ice melted. The deposit is left by the melting of a glacier where soil had been 'ploughed' along by the preceding advance of the ice sheet. Moraines comprise unstratified soil of varying composition.

Mud

It's soil , soil + water = mud

Municipal waste

Includes commercial waste and waste generated by households, with typical fractions common for all countries, although proportion depends on the source (commercial versus residential), the time of the year, as well as packaging consumption.

Natural recycling

Recycling using natural cycles including e.g. composting where organic material is returned to the soil, or use of sludge from waste water treatment plants to fertilize fields.

Nitrification

Nitrification is the process in which the nitrogen cycle works. Ammonia is created by urea and decomposition. Ammonia is turned into nitrites by nitrosamines bacteria. Nitrites are then converted to nitrates by nitrobacteria. Nitrates are used as an agricultural fertiliser. Applied incorrectly, nitrates can pose a pollution risk.

Nuclear waste

Radioactive waste produced by the nuclear industry, including waste from all stages of uranium mining and processing, as well as radioactive waste produced in e.g. medical, research and industrial activities using radioactive materials.

On-site containment

Prevention of migration of contaminants from contaminated soil to groundwater, surface water or other sensitive receptors through the construction of impermeable barriers, such as grout curtains, slurry walls and sheet-piling cut-off walls, or diversion of surface and ground water from the contaminant source area

Organic soils

Soil with a mixture of partially degraded plant & animal material and matter produced by microorganisms, allowing agriculture and other types of bio production: most soils have less than 10% organic matter, they are so-called mineral soils

Pan

A pan is a well-defined layer forming in the soil. There are two common types; a 'plough pan' which builds up in-field just below plough depth, and an 'iron pan' which forms naturally by iron oxide accumulations deposited in acid gley soils. Pans can impede the passage of water through the soil which in an agricultural context can lead to problems, if uncorrected.

Parent material

Soil parent material refers to the rocks which were weathered to form the soil in the first place. Usually the parent material is below the soil, but it can be distant if glaciers translocated the soils during the ice ages. Parent material is the focus of the study of geology.

Peat

Peat is a type of soil formed in water-logged conditions from incompletely decomposed plant material. Peat forms in wetlands or peatlands, also commonly called bogs, moors, mires, swamps and fens.

Pedology

The science of studying soils and their interaction with the wider environment.

Pedon

A three-dimensional body of soil with lateral dimensions large enough to include representative variations in the shape and relation of horizons & in soil composition. Usually 1 to $10m^2$ in area depending on variability.

Perennial plants

Perennial plants like trees are 'hardy' and can continue their growth from year to year. The opposite is an annual plant that grows, flowers and dies in just one year, an 'annual'.

Permanent Wilting Point (PWP)

The water content a soil reaches at the time when water extraction by plants has ceased as they wilt beyond recovery. PWP depends on the soil, the plant and the climate.

Pesticide

A pesticide is a chemical applied to plants (such as farm crops, or garden vegetables) to eradicate pests which would otherwise spoil the crop. Pesticides have permitted an increase in agricultural production and consistency of produce. However, applied incorrectly pesticides can pollute water bodies & aquifers and even foodstuffs can retain a minute residue of pesticides.

pH

pH stands for Potenz Hydrogen. It is measured from 1 (acid) through 7 (neutral) to 14 (alkaline), expressed on a logarithmic scale. Most soils have pH 3 to 8. Note the case of pH, a lower case 'p' and an upper case 'H'.

Photosynthesis

Photosynthesis is the important process in plant cells in which the sun's energy is used to join carbon dioxide and water to make glucose (a type of sugar), the food of green plants.

Phytoremediation

Method using plants to remediate inorganic or organic contaminants at the root zone in soils, to achieve soil stabilisation, minimisation of contaminant leaching and aesthetic improvement at low capital costs, often efficient for heavy metal removal.

Podsol

A Podsol, or Podzol, is a soil with a profile having distinctive horizons of a bleached lower topsoil, often having a cemented iron oxide 'pan' in the subsoil. Podsols are common in heathland and coniferous, or boreal, forests where there are temperate to cold moist climates.

Pores

A soil pore is the hole in-between particles of soil that becomes filled with air or water.

Precipitation

Precipitation means water reaching the ground from both rainfall, snow and hail.

Preferential flow

If the soil is cracked and fissured, then the water can pass quickly down these 'preferential' pathways. Cracks can form in drying clay soils, or beside plant roots. When rain falls to the ground, some water evaporates away again, or is taken up by plants, and some water flows off the surface towards the river. Some, however, seeps down through the soil towards the water table. Normally this water soaks through the soil and this can be quite slow depending on the soil type.

Profile

The soil profile is a vertical column of soil, essentially three dimensional and being large enough in dimension to be used to characterise the soil condition at a particular place.

Respiration

Respiration involves the exchange of oxygen and carbon dioxide between an organism and its external environment.

Runoff

Runoff occurs as water falling as precipitation does not soak deep into the soil, but passes across the surface and near-surface towards the rivers.

Saline soil

Soil subject to salinization, where the soil contains sufficient soluble salts to interfere with plant growth.

Salinization

Sometimes underground water, pumped up for irrigation, can contain mineral salts. When the water is sprayed onto the land the water is used or evaporates, but the salt can remain. Over time soil salt levels can increase - and this can be toxic to the crop plants.

Sand

That mineral fraction of the soil with particles from 0.063-2.0 mm diameter.

fine sand: 0.063-0.212 mm diameter

medium sand: 0.212-0.6 mm diameter

coarse sand: 0.6-2.0 mm diameter

Sediment

Sediment is a deposit of 'alluvium' laid down in water (such as lakes or the sea). Sediments can, over time, form rocks such as chalk and limestone.

Silt

That mineral fraction of the soil with particles from 0.002-0.063 mm diameter

fine silt: 0.002-0.006 mm diameter

medium silt: 0.006-0.02 mm diameter

coarse silt: 0.02-0.063 mm diameter

Smectites

Smectites are mineral particles in clay soils. Smectites are disc-like in shape and can slide across one-another. 'Smectitic' clays are associated with shrinking and swelling that can cause foundation subsidence in buildings.

Sodic soil

A soil with sufficient exchangeable sodium (alkali) to interfere with plant growth and cause dispersion & swelling of clay minerals.

Soil

Soil is a combination of four constituents: mineral material (sand, silt, clay and rock particles), organic material, air and water. Soil is made from the breaking down of rocks and organic matter by physical, chemical & biological weathering processes.

Soil acidity

Most soils have a pH from about 5.5 to 8, still a pretty big range - but pH 3 is really acid. Remember the smaller the pH number, the acider it is.

Soil components

Parts of soil, namely inorganic matter, organic matter, soil water (groundwater), and soil atmosphere, in a mixture; different soil types have very different amounts of these four components.

Soil heating

Commonly used thermal process for the treatment of soil with organic wastes, through heating to temperatures as high as 650 ^{0}C to achieve desorption of the organic waste from the soil matrix.

Soil mechanics

The study of how soil behaves when it is built on and great pressures build up on it. Does the soil compress, slip away or is it stable? How does the weather affect the soil?

Soil minerals

These refer to the trace elements found in soil - also called nutrients.

Soil protection

Legal protection of soil *quantity*, regulated by the acts dealing with regional planning & protection of landscape, forest, pastures, farmlands or wetlands, and soil *quality* needed to protect drinking water & the quality of food produced on the soil.

Soil remediation

Activities to remove or reduce the pollution in soil to eliminate exposure of humans and other living organisms to the contaminants and eliminate the source of contamination to ground water; remediation of contaminated soils is normally a difficult and expensive operation; remediation activities include the removal of the contaminated material through excavation or dredging and its safe off-site disposal, and on-site treatment such as destruction or containment.

Soil washing

Also called chemical extraction and soil flushing, methods in which chemicals are directly injected in the ground to remove the contaminants.

Solid waste

Any movable material that is perceived to be of no further use and that is permanently discarded; the material defined as waste depends on the management system and thus the amount of waste is often decreases as waste management develops.

Stabilization

Also called solidification, treatment of contaminated soil to contain contaminants and minimize their release to the environment.

Stomata

A tiny pore opening in the epidermis of a plant. Stomata are present in large numbers in leaves where water vapour exchange takes place (transpiration). Stomata can be opened and closed by the plant.

Substrate

An inclusive term for the soil used when describing, for instance, the portion in which plant roots exist.

Temperate climates

Temperate regions are those not experiencing very hot or very cold conditions for long periods of the year.

Texture

The description of the balance in the soil between the constituents including sand, silt and clay as well as organic matter. With experience, texture can be established accurately by touch.

Topography

The physical shape of the landscape - hills, valleys, ridges etc.

Topsoil

Topsoil is the surface layer of soil containing partly decomposed organic debris, and which is usually high in nutrients, containing many seeds, and is rich in fungal mycorrhizae. Topsoil is usually of a dark colour due to the 'organic matter' present. In arable land, 'topsoil' refers to the soil down to plough depth.

Transpiration

The process whereby plants lose water by evaporation of liquid water at the surface of the stomatal cells, the water vapour diffusing out through the leaf via the stomatal openings.

Waste collection

Separate collection of different waste fractions as a first step towards good waste management; municipalities are obliged to organise waste collection from households.

Waste management

Strategies and methods to reuse, recycle or get rid of refuse; today a main goal is to treat waste in a way which could save as much resources as possible, to avoid loss of natural resources, and see that waste is right resources in the wrong place.

Water holding capacity

Can the soil hold lots of water like a sponge? If so it has a large 'water holding capacity'. Soil organic matter increases the water holding capacity. Pure running sand has a low water holding capacity.

Waterlogged soil

A waterlogged soil is wet, with lots of water in the pores of the soil structure. The opposite is a dry aerated soil.

Weathering

The process by which materials are broken down into smaller parts and ultimately their constituents. An example is 'freeze thaw' expansion and cracking. There are both physical and chemical weathering processes.

Weed killer

A weed killer is a type of 'herbicide' used to eradicate certain types of unwanted plants or vegetation. Weed killers can be specific to certain types of plants (a 'selective weed killer').

Elements of Land/Soil Pollution

OM PRAKASH GUPTA

FEEDBACK FORM

1. Which topics from your examination point of view are inadequately or not covered in this book?

____

____

2. Is there any misprints/ mistakes/ factual inaccuracy in this book? If so, please specify.

____

____

3. What is your assessment of this book as regards the presentation of the subject-matter, expression, precision and price in relation to the other recommended books available on this subject?

____

____

4. Which competing books you regard as better than this? Please mention their author and publishers.

____

____

5. Any further suggestions/comments you would like to make for the improvement of the book?

____

____

PS: If needed, please attach a separate sheet.

Name : ____________________

College : ____________________

Address : ____________________

Email : ____________________

Mobile : ____________________

Please mail the feedback form at:

KHANNA BOOK PUBLISHING CO. (P) LTD.

4C/4344, Ansari Road, Daryaganj, New Delhi-110002

You can also email this feedback form at contact@khannabooks.com or Whatsapp at +91 – 99109 09320

BEST FEEDBACK WILL BE REWARDED SUITABLY